OPERATIONS MANAGEMENT
A Modern Approach

OPERATIONS MANAGEMENT
A Modern Approach

Rae Simons

Small-business owner and author,
Vestal, New York, U.S.A.

Apple Academic Press

Operations Management: A Modern Approach

First Published in the Canada, 2011
Apple Academic Press Inc.
3333 Mistwell Crescent
Oakville, ON L6L 0A2
Tel. : (888) 241-2035
Fax: (866) 222-9549
E-mail: info@appleacademicpress.com
www.appleacademicpress.com

The full-color tables, figures, diagrams, and images in this book may be viewed at www.appleacademicpress.com

---

First issued in paperback 2021

ISBN 13: 978-1-77463-227-7 (pbk)
ISBN 13: 978-1-926692-90-6 (hbk)

Rae Simons

Cover Design: Psqua

Library and Archives Canada Cataloguing in Publication Data
CIP Data on file with the Library and Archives Canada

CONTENTS

INTRODUCTION

Management in business pursuits and organizations can be defined as the process of coordinating people and processes in order to accomplish specific goals and objectives. Business management, specifically, includes planning, organizing, staffing, directing operations, and resourcing. Resourcing, furthermore, requires the strategic use of human, financial, technological, and natural resources in accomplishing the goals of a business entity. Because businesses can be usefully viewed as "systems," business management can be seen as human action designed to facilitate useful outcomes from that system. This view allows for the opportunity to "manage" oneself, an important factor to consider before managing operations and employees.

Business management is generally viewed as equivalent to "business administration". College and university departments that teach business management are often called "business schools," such as the Harvard Business School. Others use the broader term "schools of management" (such as the Yale School of Management), which by definition include the management of entities outside of commerce, such as in nonprofit organizations and in the public sector.

In this book, current research in one of the most important and basic aspects of business management, operations management, has been appropriately emphasized.

Business operations are those ongoing and recurring activities involved in the running of a business for the purpose of producing value for the individual owners of the business and their stockholders. The goal of successful business operations is the harvesting of value from assets owned or produced by a business. Assets can be either physical or intangible. An example of value derived from a physical asset is the profit a business makes on goods that it manufactures. An example of value derived from an intangible asset is profit derived from selling a service, such as the financial services of a bank, or a royalty derived from the licensing of a patent. The effort involved in "harvesting" this value is what constitutes business operations.

There are three fundamental goals in business operation management:

1. To secure the income and value of the business
2. To generate sustainable income over the long run
3. To increase the value of business assets

The competitive nature of twenty-first-century global commerce requires that businesses be managed strategically by managers who are knowledgeable in the principles of operations management.

— **Rae Simons**

Managing Innovation as Communicative Processes: A Case of Subsea Technology R&D

Tone Merethe, Berg Aasen and Stig Johannessen

ABSTRACT

Exploration of the communicative nature of innovation processes and the impact of this on innovation management has largely been ignored in innovation research. This paper suggests that the adoption of a complex responsive processes perspective opens up for insight and implications that depart from the prevailing view of what it means to manage joint efforts for innovation in business organizations. A key contribution is the suggestion that a change of perspective on organizations from conceptions of "whole" to notions of joint human interaction implies a need to increase management attention to the detail of local interaction between people striving to construct meaning out of new and ongoing themes, for the company and for them. We base our claims on an empirical example, drawn on a longitudinal research initiative conducted in cooperation with the Norwegian petroleum company Statoil.

Keywords: complex responsive processes, innovation management, innovation processes, subsea increased oil recovery

Introduction

Research on the phenomenon of innovation began to grow and proliferate in the 1960s, but did not truly gain momentum until about 20 years ago. Today, innovation is seen as the main enabler of long-term company viability, and is broadly recognized as being about thinking "outside the box" (Borgelt & Falk, 2007). The comprehensive interest in understanding the innovation journey (Van de Ven, Polley, Garud & Venkataraman, 1999) has been accompanied by a concurrent interest in identifying the managerial moves necessary to ensure safe arrival at a predetermined destination (Davila, Epstein, & Shelton. 2006; Ettlie, 2006; Goffin & Mitchell, 2005; Snyder & Duarte, 2003; Tidd, Bessant, & Pavitt, 2005). The results are ambiguous, but the apparent challenge of innovation management is to create an environment of perpetual innovation, where everyone is committed to excellence, resulting in growth and sustained competitive advantage. Another remaining idea seems to be that properly informed managers will be able to control the progress of innovation in such a way that the results will be, within defined limits, in accordance with some strategic intent.

This paper seeks to progress our understanding of what it may mean to manage innovation processes in business organizations. Drawing on a quadrennial research collaboration with the Norwegian petroleum company Statoil, and adopting the theory of complex responsive processes (Griffin, 2002; Shaw, 2002; Stacey, Griffin, & Shaw, 2000; Stacey, 2001; 2007), we have suggested elsewhere that the fundamental nature of innovation is communicative interaction, leading to evolving patterns of themes experienced as unpredictable and uncontrollable (Johannessen & Aasen, 2007). We have further argued that the communicative interaction can be seen as joint patterning processes of power and identity, influenced by everybody involved, although certain individuals always have a larger say (Aasen & Johannessen, 2007).

Our view of innovation as emergent patterns of themes evolving in the interplay between interdependent individuals has exposed a problem concerning established ideas of innovation management. From a complex responsive processes perspective organizational processes are joint human interaction, where individual and organizational characteristics evolve as two aspects of the same process.

Human interaction is not seen to lead to any process-independent "system," only to further communicative interaction. The management of such processes in organizations has been referred to as an activity "emerging in groups of interacting individuals engaged in collaborative action" (Tobin, 2005, p.67). This view brings to the fore a question whether one should expect individual managers to be able to design or even intentionally articulate a jointly desired generalized pattern, such as future organizational states or guidelines for collaborative behaviour.

In this paper, we discuss some of our experiences as participants in a comprehensive Statoil R&D initiative for increased oil recovery, emphasizing what we see to be important, yet traditionally overlooked aspects of innovation management. Our key findings, which are discussed in section 6, are the consequence of purposive reflection on own experiences in Statoil as well as on accounts of various events given by company members. Before we go into the discussion, we provide a brief review of the established literature on innovation management, and of the distinguishing features and the methodological orientation of the complex responsive processes perspective. Next, we will present the specific research approach, and the Subsea Increased Oil Recovery case. In conclusion, we outline some of the possible practical implications of our findings, and suggest a direction for further research.

Perspectives on Innovation Management

Industrial leaders encourage innovation based on expectations of improved business performance. Nevertheless, such processes are more often than not met with opposition, possibly because development and adoption of novelty also involves risk. Examination of various contributions within the field of innovation management leaves an impression that the main challenge of innovation management is the simultaneous handling of demands on profitability, seen as a necessity for company short-term survival, involving cost control, workforce reduction, efficiency, and value-chain optimizing; and innovation, seen as essential for long-term viability, involving creativity, experimentation, uncertainty, and the risk of failure (e.g. Ettlie, 2006; Tidd et al.; 2005; Trott, 2005). The complexity of this challenge is emphasized among others by Tidd (2001), who has made a comprehensive review of current research on innovation. He points out the random unpredictability of innovation and the diversity of research approaches as the main reasons that knowledge about innovation management still appears to be incoherent and difficult to translate into clear prescriptions.

We find that recent literature on innovation management brings into focus three particular areas of management responsibility, which are organization, competition and value realization. The main objective of organization focused innovation research is to indentify organizational characteristics promoting company innovativeness (e.g. Arad, Hanson, & Schneider, 1997; Ravichandran, 2000; Siguaw, Simpson, & Enz, 2006). The intention is to generate knowledge which managers can implement into their organization so as to increase general innovative capacity. Researchers focusing on competitive conditions analyze decisions seen to be of strategic importance, cooperation and alliances, selection of markets and market strategies, and areas for innovation. This research involves the view that managers can choose a strategic approach to innovation dependent on available resources and the competitive context (e.g. Grant, 1991; Teece, Pisano, & Shuen, 1997).

The third area, which we refer to as value realization, includes research focusing on factors having impact on the outcome of innovation processes (e.g. Durand, 2004; Neely, Fillipini, Forza, Vinelli, & Hii, 2001). In this context, organizations can be seen as actors which create and take ownership of value (Wijnberg, 2004). The realization of value as the outcome of innovation processes is related to the ability of a company to convert new knowledge, scientific breakthroughs, and technological advances into economic success. This view has engendered vast interest in theories of organizational learning (Nonaka & Takeuchi, 1995), collective knowledge (Glynn, 1996), knowledge management (Quinn, Anderson, & Finkelstein, 1998), communities of practice (Wenger, 1998), and indeed, innovation management (Davila et al., 2006; Tidd et al., 2005; Trott, 2005). Furthermore, the effect of collaborative processes on innovation and business performance is discussed by several researchers (Cohen & Levinthal, 1990; Durand, 2004; Tsai, 2001; Chesbrough, Vanhaverbeke, & West, 2006).

Although the defining feature of processes of innovation is pointed out to be complexity and uncertainty (Tidd, 2001), most authors hold on to assumptions of management controllability of such processes, justified by the observation that many companies survive and renew over time (Tidd et al., 2005). As distinct from most researchers, Van de Ven et al. (1999) point out that managers at many hierarchical levels are involved in the management of innovation, and that in spite of a widespread view that managers have a uniform, common perspective; managing innovation does involve diversity and conflict. Accordingly, their suggestion is that innovation processes may be inherently uncontrollable and that a relaxation of "traditional notions of managerial control" (p. 66) is needed. In this paper we address the problem described by Van de Ven et al. (ibid.), by adopting a complex responsive processes perspective as the theoretical basis.

The Complex Responsive Processes Perspective – Distinguishing Features and Methodological Implications

As will be further described in section 4 and 5, our research situation was highly participative. Implicitly, a theoretical perspective focusing on "real time" human interaction was demanded. The basic idea of the complex responsive processes perspective, which is the perspective adopted in this study, is the importance of taking the experiences of human action and interaction seriously (Stacey, 2001; 2007). The distinguishing features of this perspective are that all human relating is seen as fundamentally communicative, and that ideas of the autonomous individual and the objective observer/manager are replaced by assumptions of the simultaneous social construction of group and individual identities (Stacey, 2007). The methodological position is that of reflexivity in both individual and social terms.

An important source of inspiration of this perspective is Elias (1978; 2000). He argued that even if people do plan and do have intentions, they inevitably also participate in figurations of interaction in which power is always an intrinsic property. He saw such figurations of power to be asymmetric, meaning that individuals influence the communicative interaction they are part of to a greater or lesser extent. Elias therefore suggested a non-linear, paradoxical, transformative causality between human action and the outcome of social processes. The non-linearity means that even small variations in themes have the potential to lead to radical global re-patterning of conversations and power relations, sometimes referred to as the "butterfly effect" (Lorentz, 2000). For organizations, this should mean that conflicts between actions, plans and purposes of interdependent people leads to repetition and change as aspects of the same process, and that this is essential for novelty to emerge and evolve (Leana & Barry, 2000). Elias' view could therefore be seen as an opportune reminder that human nature makes it improbable that one individuals' plan or intention should become dominant as the long-term reality of everybody in the organization.

The methodological orientation of the complex responsive processes perspective is grounded on the thinking of George H. Mead (Stacey & Griffin, 2005). Mead saw human social life as perpetual movement, emerging in responsive processes of negotiation and reality constructions. He argued that mind, consciousness, self-consciousness and society are reflexively co-constituted through communicative gesture and response cycles of meaning-making, in which individuals relate and act in cooperative-conflictual interdependencies with other people (Mead, 1967).

Elias and Mead shared the view that reality is seen to develop because of social interaction. In consequence, the source of creativity and innovation lies in the transformative potential of the continuous mutual adjustments of meaning between people doing whatever they do during their working days. Note that from this perspective people are always seen to be in practice, and practice is experienced through participation. Furthermore, experience is understood as personal experience of interaction. Consistent with these ideas, the present study was carried out with an explorative attitude referred to as emergent participative exploration (Christensen, 2005). The term emergent is conceptualized to represent the formation of meaning for the participating researcher from the exploration of one activity or situation. The emerging meaning then guides the suggestion of the next activity of exploration, as well as the development of specific themes appearing to be of particular relevance to the situations explored.

Characteristic of this approach is the connection made between research and identity, implying the view that researchers, as participants in organizational everyday life, influence the processes studied. Even more importantly, it is assumed that the researcher may change in the same process. Emergent participative exploration therefore shows similarity, but also distinct differences, to action research (discussed e.g. by Williams, 2005). In our view, emergent participative exploration can be seen as the researchers' intent of experiencing everyday social processes in organizations, rather than being about a particular method. This means that we see it appropriate to apply different qualitative methods as we know them traditionally, as part of the process of developing new meaning.

Research Approach

In connection with the establishment of six strategic research and development (R&D) programs in Statoil in 2003/2004, an idea had developed among a few managers in the Technology division that one of the programs, called Subsea Increased Oil Recovery (SIOR), provided a good starting point for evaluation of their internal processes for innovation. The four members of the SIOR core team (CT) accepted the presence of a researcher as part of their team, and so the first author of this paper was invited to join them. She was engaged in Statoil for four years, and granted an ID card, the opportunity to work on-site, and to access internal databases, e-mail system and intranet. The study lasted from January 2004 to October 2007, but the collaboration with StatoilHydro is still ongoing.

The study mainly involved participation in ongoing formal and informal meetings and conversations intended to lead to innovation in Statoil, but also interviews, consultative intervention and studies of written material. As an implication of the research situation, our view of innovation management in Statoil is primarily based on experience from the Research Centre. We were, however, given the opportunity to interview leaders and workers in all the business areas involved: Technology & Projects, Exploration & Production Norway, and International Exploration & Production, many of them holding key positions. This gave us the opportunity to evolve and challenge our impressions of ongoing company activities for change and innovation, including managers' contributions to these processes. The respondents interviewed were persons recommended by our main contact persons in Statoil. In addition, we included persons we met during the study who, assumingly, would help us broaden our perspective on Statoil innovation processes.

Examples from the SIOR program are described and analyzed by use of a narrative style. The narratives are based on three kinds of input, which are the factual story about evolving events, individual stories offered in conversations between the first author and Statoil members, and finally, own reflections about experiences as SIOR program participant. The SIOR program made a very rich case, and in the process of going through notes and memories, we intentionally picked out and composed stories to support the communication of our present understanding of the innovation processes we were part of. Our analysis could therefore be seen to be our understanding of the particular and general themes discussed by people we spoke to. A key objection to this research, as it may be to all participative research methods, is thus the biased subjectivity of the approach by which the results are obtained and presented. In line with Peller (1987), we argue that in the end, research results always depend on the subjective choices of the researcher about what to do, how to do it, and how to interpret, analyse, and present the material. The analytical method used is that of intentionally reflecting on the details of one's experience of organizational processes, as basis for new insights and practices.

Statoil and the Sior Case

The Norwegian oil and gas company StatoilHydro is the leading operator on the Norwegian continental shelf (NCS). With its modest 29,000 employees (current numbers), it is a relatively small company compared to its competitors. The conditions on the NCS have, however, made extreme demands on technology. This has lead to the recognition of the company as a world-leader in the use of innovative technology, and brought it to the position of being the world's largest operator of deepwater fields. According to their web-site (www.statoilhydro.com), StatoilHydro aim at sustaining long term profitable growth through increased international activities and renewed efforts on the NCS. The latter scope is connected to an objective to increase oil recovery, but also to a necessity of avoiding the hidden threat of having to close down fields which are becoming unprofitable.

Compared to the previous, comprehensive field developments, recent NCS fields are smaller, and present value estimates leave little time and money for technology development. Still, the need for innovation is as pressing as ever. This has promoted a demand for increased collaboration within the company to develop new technology and work processes across business assets. As most StatoilHydro projects and operations are also conducted in cooperation with other companies, the management of development activities in the company by and large implies managing collaborative teams composed of people from various internal departments, and from one or more external companies.

The strategic R&D programs initiated in the Statoil Research Centre 2003/2004 were seen to be an important measure to promote the renewed efforts on the NCS. These were umbrella programs intended to embrace and adapt ongoing research activities according to program ambitions, as well as to frame new initiatives. The SIOR program was granted an annual budget of about 25 million Euros. The objective of the program was to provide technology making probable the increased production of oil from existing and future Statoil operated subsea fields on the NCS from the 2003 average of 43% to an average of 55% in 2008. This would mean the production of about 1.4 billion barrels of extra oil over the estimated lifetime of the fields, corresponding at the time to an added gross profit of about 70 billion US$. This ambition was judged to be unattainable through the use of existing technologies, and implied the need for accelerated development and testing of technologies in the pipeline, as well as for the generation of completely new concepts.

As no single concept could meet the 55 % SIOR ambition, the program eventually embraced about 25 different development activities, spanning technology connected with the identification of drainage points and intervention needs, improvement of production management, provision of low cost drainage points and intervention, reduction of well head pressures and increase of liquid handling capacity. The activities involved more than 100 persons, of whom about half were employed in other companies. A core team (CT) of four hand-picked people was composed to manage the program. All of them were experienced managers and specialists having worked many years in Statoil operational units, and representing different disciplines within petroleum engineering.

Few SIOR activities were started from scratch. Work had been ongoing in the Research Centre and in other parts of Statoil for a long time, aiming at the development of various technologies to render possible the recovery of increased volumes of oil. The experience of those involved, among them future members of the SIOR core team, was that it was hard to attract attention towards this kind of initiatives in the company, and that it generally did not result in anything. The idea of framing individual projects having similar intentions into an umbrella program with an overall ambition was therefore launched as an attempt to direct the top management's attention towards the potentiality of such activities, and to facilitate communication about them in the company. The approach appeared to work, as the SIOR ambition gradually got fully backed up by the top management.

Innovation Management as Acts of Participation

The basic idea of the SIOR program was that the combination of operational unit requirements and the creative capabilities of Statoil researchers and specialists, and experts from other companies, would lead to the generation of new technologies tailored to enable the increased efficiency of subsea oil production. Statoil was taking on the dual role of customer and technology provider, assigning the role of provider to its Research Centre. The argumentation was in line with Thamhain (2003), who claims that the hallmark of capable R&D teams is that they generate innovative ideas, but also "transfer newly created concepts through the organizational system for economic gain" (p. 297), and with von Hippel (2005), who asserts the importance of user-centred innovation. The program was characterized by the assignment of a specified, measurable end target, and the diversity of technology developments needed to succeed.

SIOR was, however, also about the introduction and implementation of new ways of working in the company. The most important of the change initiatives was the top management expectation about closer cooperation between Statoil researchers and members of operational units. Until then most of the larger development projects had been performed in collaborations between Statoil operating unit members and external suppliers, and the role of the researchers in these projects was generally seen to be modest. The authorization given to SIOR members to take a leading role in business development was therefore seen as unusual, and was referred to as a "different way of working." In this particular case, then, innovation management was not only about managing technology development, but also about preparing for the adoption of new technology elements by Statoil-operated fields, as well as bringing it all about through approaches to collaboration which in several ways were unfamiliar to the persons involved. In this section, we address key findings related to this versatile task, seen from the perspective of complex responsive processes. The findings are illustrated in Table 1.

Table 1: Key issue and key findings of research

Key issue	Method	Key findings
Assuming that innovation is communication processes, what does it mean to manage innovation in business organizations?	Emergent participative exploration	a) The principal task of innovation management is participative actions intended to influence emerging patterns of themes, to support the collective movement towards a desirable future situation b) Innovation management involves dealing with various phenomena of group dynamics, including insider/outsider relations. Top management support and the attribution of credibility to the innovation manager from involved parties are important supportive factors. c) Innovation management is the courageous, continued exploration of experiences of being together in spite of potential conflicts. This involves the acknowledgement of paradox, and of the simultaneous need for sustained pluralism and widespread joint meaning. d) The idea of management participation is on a collision course with predominant management ideas. of planning, predictability, control, and monitoring

Managing Phenomena of Group Dynamics

A common objection among operational unit members to the leading role in innovation given to the Research Centre was that they understood better the business challenges facing Statoil than did most of the researchers. Although most acknowledged that there were a lot of excellent niche experts in the Research Centre, operational unit members also meant that they were better trained as project managers. From their side, several of the researchers expressed a view that neither members of the SIOR CT, nor of the operational units, fully appreciated that inventing technology was somewhat more complicated than to "sit down and just decide to get a break-through." In some cases, this seemed to lead to a clash of interests which constrained cooperation between members of the operational units, and the researchers. However, the full support given to the SIOR program by the top management appeared to pressure the involved parties to exert themselves so that activities progressed.

In line with Van de Ven et al. (1999) we observed that the differences between the members of the Research Centre and the operational units were also reflected among the managers. Partly because of this, the role of the SIOR CT members gradually evolved to be that of technology broker, or intermediary between development and adoption processes. We were told that this was a new role in the company, judged to be very important for the realization of the SIOR ambition. The SIOR CT members' knowledge of current business processes appeared to provide the credibility among operational unit managers that was not attributed to many of the researchers. The reputation of the head of the SIOR CT as an experienced and able manager was particularly important, and opened doors to persons and meetings of great importance to SIOR. On the other hand, from our position is seemed that the former experience of the SIOR CT members made them somewhat less accepted by members of the Research Centre.

The preceding elaboration suggests that research attention should be directed to the theme of identity in groups, experiences of inclusion and exclusion, and the quality of relations. Mead (1967) claimed that "we" identities in groups are based on generalisations. As "we" identities develop, simultaneous perceptions of "them" evolve, resulting in a paradoxical dynamics of inclusion and exclusion of individuals. Such activities, which will always be part of the process of forming and identifying with a group, are accompanied by the tendency to label groups of people in ways that enforce the difference between "us" and "them" (Stacey, 2007). This tends to leave us with an impression of the uniformity of group characteristics. Hence, what people produce when they interact in the living present could be seen as the continuous creation and recreation of individual-group identity. In line with our experience, Elias and Scotson (1994) argue that raising issues of "us" and "them" relationships can uncover insider and outsider relations in which diverse groupings develop or damage their cooperation. Correspondingly, Stacey (2007) points out that power differentials between groups may create a powerful dynamics in organisations, probably constituting one of the main reasons for the failure of attempts to realise strategic intents. Seen like this, the task of the SIOR CT members could be seen to be the re-patterning of the experience of identity among people involved in SIOR, with the intention to enable new ways of working together.

Managing Emerging Patterns of Themes

The SIOR CT members were continually participating in meetings and conversations, sometimes involving two persons, sometimes twenty. They kept repeating the 55 % ambition, but at the same time they spoke of specific technology elements, such as Light Well Intervention; Wet gas compression; or Shared Earth modelling, and about the business opportunities related to the adoption of such technologies by the various NCS fields. In concurrence with this, they made contact with Statoil information associates, who they educated by inviting them to relevant expositions and meetings. They also made sure that the top managers were always provided with the latest SIOR presentation material. After about two years it was commented in a core team meeting that they had been talking so much about the SIOR technologies that people in other petroleum companies started to implement them before Statoil did. Gradually, what emerged in the wake of this intensive communicative effort was a widespread opinion in the company that the 55 % ambition and the activities performed as part of the SIOR program were appropriate.

Our finding is that the ideas of the SIOR program was formed by various interests, but at the same time gradually formed, and transformed, the interests of those who in some way were engaged in the activities. This is in accordance with a perspective on organizations as patterns of interactions, where ongoing processes are influenced by many individuals in many roles, deliberately and unconsciously. It also supports one of the core ideas of the complex responsive perspective, which is that meaning is not determined by a gesture (like a statement or a move), but by the responses brought out by that gesture (Stacey, 2005). This leads to a cyclic movement of further gestures and responses in which meaning is formed, and power relations and senses of identity and meaning potentially affected. Implicitly, when people attempt to design or change some global pattern, like "us" and "them" relationships, they are doing nothing more than making a gesture, although this can be a very powerful gesture (Elias, 2000). What is emerging, are patterns of responses that no one individual could have decided, and that may involve the experience of novelty. It is, however, not given that the emerging patterns are perceived as "suitable" or "successful." We see this as a particularly important point, which is rarely taken into account when innovation processes are discussed.

This further underlines a problem attached with the idea that innovation can be predetermined by the actions of particular individuals, such as managers, and emphasizes the intrinsic collective nature of innovation processes. At the same time, it indicates that individual and joint experiences of meaning and identity are found, and alters, in the ordinary, everyday conversations between people at work. In our view, this emphasizes the need to increase management attention on the detail of local interaction between people striving to particularize the significance of new and ongoing themes for the company and for them. Accordingly, the present rather one-sided focus on management acts as the development and following-up of steering documents and key performance indicators should be replaced by the recognition of the potentially even greater significance of management as acts of participation. Two points should be attached to this claim. One is that, as participants in the social processes of organized life, managers should be regarded as free, and at the same time constrained, in choosing their own actions. Secondly, the actions they decide will expose their colleagues and subordinates to both possibilities and constraints.

Innovation Management as Disturbance and Stabilization

The enabling constraints intrinsic in all human relating and the dynamics of inclusion/exclusion created between groups indicate that paradox is an inevitable part of such everyday interaction. We see the experience of paradox as being of particular importance in relation to innovation, which inherently entails the introduction or emergence of novel ideas. This means that reproduction of currently stabilized themes is disturbed, and so it is reasonable to expect that novel ideas may be perceived as controversial. This may cause uncertainty and even conflict between individuals and groups who incline towards re-establishing habitual patterns of themes, and those who pursue the new ones. Shifting experiences of identity and difference, inclusion and exclusion, inspiration and anxiety, freedom and control, and of structures of power, are likely to cause enthusiasm with some, and doubt and resistance with others. Such processes should not be seen to approach a mature or final state, but as being continued by individuals participating in local, everyday interaction, perpetually creating and recreating ideas about their intentions and possibilities.

While much of the existing management literature is about punctuating paradoxical situations, for example by recommending the introduction of unambiguous, measurable ambitions and objectives, our view is that managers should rather seek to capitalize on the pluralism which is the inherent property of paradox. Innovation can be found nowhere but in the emerging patterns of themes, which spread out and evolve because people take part in many local conversations. On the other hand, the intrinsic challenge of organizational achievement is the need for coordinated action between many people, and such collaboration depends on the production of "emergent, coherent, meaningful patterns of interaction both locally and population-wide at the same time" (Stacey, 2007, p. 434). Our findings indicate that in the myriad of ordinary, local meetings, themes are reiterated and sometimes transformed. This may enable joint action, or inhibit it. We therefore suggest that innovation depends not only upon the emergence of novel patterns of themes, but also on the diffusion and temporary stabilization of the evolving patterns among people whose cooperation is needed to render possible the enactment of the new themes.

Consistent with this idea, the continued intention of the SIOR CT members was clearly to bring locally evolving patterns of themes under the sway of the SIOR ideas in such a way that the decisions made by field directors or others whose actions affected the program activities were in favour of the realization of the SIOR ambition. A problem frequently discussed was how to ensure this. As expressed by one of the CT members: "The problem is not the conversations I am part of, the problem is those conversations in which I am not a part." His experience was that patterns of talk emerging in his presence did indeed change as they were further evolved in conversations between other people, and not always in ways seen by him as favourable.

According to Fonseca (2002) pluralism in conversation is of vital importance for innovation. He further suggests that this pluralism is experienced by humans as misunderstanding. By misunderstanding he seems to mean the lack of joint meaning, leading to the continual shift and evolvement of the patterning processes of new themes because of the current introduction of new themes and ideas into the emerging patterns. Fonseca argues that the continued disturbance of emerging thematic patterns of experience may prevent premature or unwanted stabilization of themes. We see this interpretation of misunderstanding as a support to our observation that innovation emerged from prolonged communicative processes characterized by conflict, ambiguity and persuasion. In consequence, innovation management involves the courageous, continued exploration of the experiences of being together in spite of potential conflicts. On the other hand, tolerance for the kind of misunderstanding described by Fonseca (2002) appeared to be relatively low in meetings between SIOR members and people working in the operational units. A widespread expectation in Statoil seemed rather to be that the particular intention of the SIOR CT should be the enabling of controlled movement towards a desirable future organizational state.

Another experience was that the frequent introduction of new themes made by the Statoil top management and by other people in managerial positions represented a diversion of attention in the company from the SIOR ambition towards competing tasks and ideas. In consequence, the re-stabilization of the SIOR idea through the repeated communication of possibilities and promising results appeared to be an aspect as important for innovation success as was the maintenance of ambiguity.

A Problem with Participation

Prevailing management values, placing emphasis on efficiency, monitoring, control and short-term profit (Miles, 2007), seemed to be highly esteemed in Statoil, even in connection with processes referred to as innovation. In addition, the top management appeared to be inspired by the principles of value based management (Black, Wright, Bachman, Makall, & Wright, 1998). Among other things, this implied the introduction of prescriptive statements and visionary themes intended to direct the attention of company members towards specific objectives, like the need for innovation. Given the current view of the significance of innovation for business prosperity, it is no surprise that people in executive positions are prone to the temptation of subjecting innovation processes to the same procedures for strategy and control as other business processes. Moreover, the dominance of research ignoring phenomena of group dynamics and non-linear, time dependent effects of action involves that managers are offered models of organizational processes that do not and cannot capture the temporally embedded accounts that enable them to understand how emerging and evolving patterns come to be.

The current management expectation about performance measurements appears to be in opposition to our view of innovation management as the purposeful participation in communicative interaction. In Statoil, there seemed to be a never-ending demand for various kinds of project documentation. Although this was emphasized to be a problem by several of the managers engaged in SIOR activities, because it meant that they often had to leave their subordinates to their own devices, most seemed to accept the situation. Only a few openly said that they prioritized communication with their subordinates, to help them particularize the generalized project ambitions into their specific situations, and also to encourage people to communicate face to face about new ideas, within and across disciplines. "They need to see themselves in all the "us,"" one of these managers pointed out. The dilemma she and others faced when they focused on participation was that this kind of action was not "visible" in the company; their individual level of achievement could not be measured. Given the predominant management thinking in Statoil, they therefore felt that they were not really recognized as leaders by their superiors.

Paradoxically, it seemed that while what the SIOR members largely wanted was management attention towards their professional skills and challenges, the top management gradually tightened the demand for proofs of control, such as accounts and forecasts. This suggests that a question of particular interest is how decision makers think about their intentions when they are suggesting propositions and even orders, and of the possibilities and limitations of using target setting, planning and monitoring as basis for long-term organizational performance. From our perspective, the introduction of prescriptions and visionary themes are indeed important tools in the process of leading, and can be seen as factors contributing to the creation of meaning, but not as contributors to the achievement of control. Rather, mechanisms for structure and control should be seen as nonresponding, response provoking tools, causing organizational members to feel enabled, but also constrained by their implications, depending on situation.

Our experience was that the actions following the introduction of the 55 % ambition emerged as the outcome of interplay between rules, plans, intentions, and choices, and were primarily patterned as narrative themes, affected by and affecting ongoing and emerging patterns. Management attention towards the challenges and needs of persons engaged in development projects clearly stimulated and supported innovation efforts. Based on this, we see the principal task for managers of innovation to be explorative and participative actions intended to inspire and motivate, but also "force," the members of an organization towards the joint creation and realization of an imagined future, concurrently guided by the insight that the future is unknowable. Accordingly, management inclination to focus on general values and on control and monitoring of development processes could be seen as a disregard for the significance of participation as a management tool in innovation processes. The issue should rather be what purpose prescriptive and visionary formulations serve, what the emerging patterns of communication in ordinary everyday organizational life in response to such formulations are, and how we can understand innovation processes as being a part of this.

Implications of Research

Our experience in SIOR demonstrates that company members are influenced by innovation; yet will influence the processes, consciously or unconsciously, at the same time. Therefore, innovation is a collective achievement, and, implicitly, a collective responsibility. This means that innovation managers should pay attention to individuals' possibility to go on together in the face of the paradoxical dynamics of conflict and cooperation characterizing innovation processes. An issue in this is the need that managers engage in conversations about the apparent tension between demands for efficiency and for innovation, and about the influence of their own actions on ongoing and emerging innovation processes. In our view, the complex responsive processes perspective basically offers the opportunity to reflect on the manner in which people are reasoning as one of many aspects of human action in organized life, instead of taking rationality for granted. For managers, this means the possibility to take seriously their own experiences as leaders by focusing on what they actually do, and not on what they did or plan to do. Accordingly, we suggest that anagement team members may profit from spending more time in joint reflection not only about how to solve technological problems and meet with measurable business targets, but on own relevant everyday experiences, to explore what these experiences may mean to their individual and joint possibility of and capacity for innovation. Similarly, increased awareness of approaches to explore, clarify, and possibly develop emerging patterns of themes among innovation process participants appears to be a valuable investment of time.

Although preliminary, the results of the present study encourage continued exploration of the complex responsive processes perspective as an alternative way to research and explain processes intended to lead to innovation in companies. The principal objective of future research should be to extend our understanding of human interaction related to the development and exploitation of innovation in and between commercial companies and other organizations, to learn more about communicative aspects salient in innovation processes, and how these are enacted in different contexts. The objective is to understand better how such processes can be managed and supported, and to explore alternative approaches to describe innovation in terms of profitability and growth. We see individual and joint processes of particularization and functionalization of generalized ideas to be of particular research interest, because these processes are prerequisite for managers to be able to enact innovation in ordinary, local interactions in the living present. Accordingly, researchers' attention should be on issues and aspects emerging from the responsive processes of relating between people, in which thematic patterns caused among other things by considerations related to profit, politics, safety and reputation, but also to technology, physical environments, as well as habit, intertwine and evolve in unpredictable ways.

Conclusion

Drawing on a 4-year engagement in Statoil, we have discussed important aspects of innovation management, seen from a complex responsive processes perspective. Our study was carried out in accordance with the ideas of emergent participative exploration. We bring to the fore a view that innovation processes involve widespread movement of thought, which are changes in action. Implicitly, management measures of presence and support should be focused more than measures of control. Our suggestion is that innovation management is participative actions intended to influence emerging patterns of themes, to support the collective movement towards a desirable future situation. Our findings indicate, however, that the idea of management participation is on a collision course with predominant management ideas of planning, predictability, control, and monitoring.

Innovation management can be seen as the courageous, continued exploration of experiences of being together in spite of potential conflicts. This involves the acknowledgement of paradox, and of the simultaneous need for sustained pluralism and widespread joint meaning. It also involves dealing with various phenomena of group dynamics, including insider/outsider relations. We find that top management support and the attribution of credibility to the innovation manager from involved parties are important success factors.

Acknowledgements

An earlier version of this paper was presented at the ICA 2008 conference in Montreal. We thank Professor James Taylor of the University de Montreal for the valuable insights provided on that manuscript. We also thank the IJBSAM editor and peer reviewers for their helpful comments on this manuscript.

References

Aasen, T. M. B., & Johannessen, S. (2007). Exploring Innovation Processes from a Complexity Perspective. Part II: Experiences from the SIOR case, International Journal of Learning and Change, 2(4), 434–446.

Arad, S., Hanson, M. A., & Schneider, R. J. (1997). A framework for the study of relationships between organizational characteristics and organizational innovation, J. Creative Behaviour, 31(1), 42–59.

Black, A., Wright, J., Bachman, J., Makall, M., & Wright, P. (1998). In Search of Shareholder Value: Managing the Drivers of Performance, London: Pitman. Borgelt, K., & Falk, I. (2007). The leadership/management conundrum: innovation or risk

management? Leadership & Organization Development Journal, 28(2), 122–136. Chesbrough, H.; Vanhaverbeke, W., & West, J. (eds.) (2006). Open Innovation: Researching a New Paradigm. Oxford, NY: Oxford University Press Christensen, B. (2005). Emerging participative exploration: consultation as research. In: R. D. Stacey, & D. Griffin (eds.) A complexity perspective on researching organizations. Taking experience seriously (pp. 74–106). Oxon: Routledge

Cohen, W. M., & Levinthal, D. A. (1990). Absorptive capacity: A new perspective on learning and innovation. Administrative Science Quarterly, 35(1), 128–152.

Davila, T., Epstein, M. J., & Shelton, S. (2006). Making innovation work. How to manage it, measure it, and profit from it. Upper Saddle River, NJ: Wharton School Publishing.

Durand, T. (2004). The strategic management of technology and innovation. In: T. Durand, O. Granstrand, C. Herstatt, A. Nagel, D. Probert, B. Tomlin, & H. Tschirky (eds.), Bringing technology and innovation into the boardroom. Hampshire: Palgrave Macmillan, European Institute for Technology and Innovation Management.

Elias, N. (1939/2000). The civilizing process. Oxford: Blackwell.

Elias, N. (1978). What is sociology? New York: Columbia University Press.

Elias, N., & Scotson, J. (1965/1994). The established and the outsiders. London: Sage

Ettlie, J.E. (2006). Managing innovation. New technology, new products, and new services in a global economy. 2nd ed., Burlington, MA: Elsevier Butterworth-Heinemann.

Fonseca, J. (2002). Complexity and Innovation in Organisations. London: Routledge.

Glynn, M. A. (1996). Innovative genius: A framework for relating individual and organizational intelligence to innovation. Academy of management review, 21(4), 1081–1111.

Goffin, K., & Mitchell, R. (2005). Innovation management: Strategy and implementation using the pentathlon framework. New York: Palgrave Macmillan.

Grant, R. M. (1991). The resource-based theory of competitive advantage: Implications for strategy formulation. California management review, 33(3), 114–135.

Griffin, D. (2002). The emergence of leadership: Linking self-organization and ethics. London: Routledge.

Johannessen, S., & Aasen, T. M. B. (2007). Exploring Innovation Processes from a Complexity Perspective. Part I: Theoretical and Methodological approach. International Journal of Learning and Change, 2(4), 420–433.

Leana, C. R., & Barry, B. (2000). Stability and change as simultaneous experiences in organizational life. Academy of Management Review, 25(4), 753–759

Lorentz, E. N. (2000). Predictability Does the Flap of a Butterfly's Wings in Brazil Set Off a Tornado in Texas? In: R. Abraham, & Y. Ueda (eds.) The Chaos Avant-Garde: Memories of the Early Days of Chaos Theory (pp. 91–94), World Scientific, Series A, 39

Mead, G. H. (1934/1967). Mind, Self and Society. Chicago: Chicago University Press.

Miles, R. E. (2007). Innovation and Leadership Values. Californian Management Review, 50(1): 192–201.

Neely, A., Fillipini, R., Forza, C., Vinelli, A., & Hii, J. (2001). A framework for analyzing business performance, firm innovation and related contextual factors: perceptions of managers and policy makers in two European regions. Integrated Manufacturing Systems, 12(2), 114–124.

Nonaka, I., & Takeuchi, H. (1995). The knowledge creating company. New York: Oxford University Press.

Peller, G. (1987). Reason and the mob: The politics of representation. Tikkun, 2 (3), 28–95.

Quinn, J B., Anderson, P., & Finkelstein, S. (1998). Managing professional intellect. Harvard Business Review on Knowledge Management, Boston, MA: Harvard Business School Publishing.

Shaw, P. (2002). Changing conversations in organizations. A complexity approach to change. London: Routledge.

Siguaw, J. A., Simpson, P. M., & Enz, C. A. (2006). Conceptualizing innovation orientation: A framework for study and integration of innovation research. J Product Innovation Management, 23, 556–574

Snyder, N. T., & Duarte, D. L. (2003). Strategic innovation. Embedding innovation as a core competency in your organization, San Francisco, CA: Jossey-Bass.

Stacey, R. D., Griffin, D., & Shaw, P. (2000). Complexity and Management—Fad or Radical Challenge to Systems Thinking? London: Routledge.

Stacey, R. D. (2001). Complex responsive processes in organizations. Learning and knowledge creation. London: Routledge.

Stacey, R. D. and Griffin, D. (Eds.) (2005). A complexity perspective on researching organizations. Abingdon: Routledge.

Stacey, R. D. (2007). Strategic management and organisational dynamics: The challenge of complexity to ways of thinking about organisations, 5th ed. London: Prentice Hall.

Streatfield, P. J. (2001). The paradox of control in organizations. London: Routledge.

Teece, D. J., Pisano, G., & Shuen, A. (1997). Dynamic capabilities and strategic management. Strategic Management Journal, 18(7), 509–533

Tidd, J. (2001). Innovation management in context: Environment, organization and performance. Int J Management Reviews, 3(3), 169–183.

Tidd, J., Bessant, J., & Pavitt, K. (2005). Managing innovation. Integrating technological, market and organizational change, 3rd ed. Chichester: Wiley.

Tobin, J. (2005). The role of leader and the paradox of detached involvement. In: D. Griffin, & R. D. Stacey. (eds.): Complexity and the experience of leading organizations (pp. 61–92). London: Routledge.

Trott, P. (2005). Innovation management and new product development. 3rd ed. Harlow: Prentice Hall.

Tsai, W. (2001). Knowledge transfer in intraorganizational networks: Effects of network position and absorptive capacity on business unit innovation and performance. Academy of Management journal, 44(5), 996–1004.

Van de Ven, A. H., Polley, D. E., Garud, R., & Venkataraman, S. (1999). The innovation journey. New York: Oxford University Press.

von Hippel, E. (2005). Democratizing innovation. Boston, MA: MIT Press books

Wenger, E. (1998). Communities of Practice: Learning, Meaning, and Identity. New York: Cambridge University Press.

Wijnberg, N. M. (2004). Innovation and organization: Value and competition in selection systems. Organization studies, 25(8), 1413–1433.

Williams, R. (2005). Belief, truth and justification: issues of methodology, discourse and the validity of personal narratives. In: R. D. Stacey & D. Griffin (eds.): A complexity perspective on researching organizations (pp. 43–73). Abingdon: Routledge.

Deferred Action: Theoretical Model of Process Architecture Design for Emergent Business Processes

Nandish V. Patel

ABSTRACT

E-Business modelling and ebusiness systems development assumes fixed company resources, structures, and business processes. Empirical and theoretical evidence suggests that company resources and structures are emergent rather than fixed. Planning business activity in emergent contexts requires flexible ebusiness models based on better management theories and models . This paper builds and proposes a theoretical model of ebusiness systems capable of catering for emergent factors that affect business processes. Drawing on development of theories of the 'action and design'class the Theory of Deferred Action is invoked as the base theory for the theoretical model. A theoretical model of flexible process architecture is presented by identifying its core components and their relationships, and then illustrated with exemplar flexible process archi-tectures capable of responding to emergent factors. Managerial implications of the model are considered and the model's generic applicability is discussed.

Keywords: service science, Theory of Deferred Action, theoretical model, business process, process architecture, emergence

Introduction

The term process architecture describes artefactual objects resulting from design-ing and implementing business processes and the supporting IT systems. This is planned architecture. Process architecture and business organization should be in reciprocal relation to each other to improve business performance. The reciprocity is between organization and its IT systems. Research into process architecture has focused on practical aspects resulting in a body of literature on business process modelling (Miers D 1994; Elizinga D, Horak T et al. 1995; Georgakopoulos D, M et al. 1995; Reijswoud V, Mulder H et al. 1999). Less research is evident on building models based on theory, see for example Shaw et al. (2006) The question considered theoretically in this paper is how business process designers can model changing processes involving uncertainties, unpredictable futures, and non-stan-dardisable business processes. This type of business process is termed emergent business processes (EBP) and it is not amenable to planned architecture design.

A search of leading business and management journals revealed that there is little research into the theory of process design. The journals consulted were Harvard Business Review, Academy of Management Journal and Academy of Management Review for the period 1999-2006. Slack (2005) proposes a model and Larger and Horte (2005) a taxonomy indicative of emerging theoretical per-spective on process architecture design. This is in contrast to editorial support for theoretical work in management in general (Kilduff M 2007).

Editors of leading management journals and information systems journals value theory building and theory publishing (Weber R 2003). Researchers in in-formation systems have proposed theoretical models to explain IT systems de-sign for organizational knowledge management (Markus M L, Majchrzak A et al. 2002) and executive information systems (Walls J G, R et al. 1992). In business and management, there is little similar theory proposed for process architecture design. Shaw et al., (2007) propose a theory-based process architecture capable of evolving with business process change. They state that: 'there is no theoret-ical basis for any assembly of business process model constructs that we have seen.' (p.95). Kettinger and Grover (1995) propose a theory of business process management. A theoretical model of process architecture design is proposed in this paper to improve theoretical process knowledge and to address the general gap in theory building for process architecture design. The theoretical basis of process architecture design in general is weak.

Theoretical understanding provides a sound basis for designing process architecture. The central question addressed here concerns the kind of process architecture design required for emergent business processes. A theoretical understanding of process architecture design should enable appropriate responses to emergent organization—its processes and the associated process information and process knowledge. This paper addresses three related problems in process architecture design: emergent organization, planned business change, and emergent business processes (also termed non-standardisable processes). The crucial interrelationship among these problems is not addressed in the research literature.

Theoretical models can draw on two types of theory: variance theory or process theory (Soh C and Markus M L 1995). Variance theories explain the variations in the magnitude of a certain outcome and are better at explaining 'why' something happens. Variance theories are also called linear models because of the use of linear equations to model phenomena. Variance theories do not explain well situations where the outcome is uncertain—sometimes occurring, sometimes not. Such situations indicate that the necessary conditions are not sufficient to produce the outcome. Process theories better explain situations where the causal agent is not sufficient to produce the outcome. Process theories explain 'how' something happens. They are better suited to explain process architecture design for emergent organization, as the outcome is uncertain in emergent organization and the causal agent alone is insufficient to produce the outcome.

Drawing on process theory, this paper presents a theoretical model of process architecture design for emergent business processes and emergent organization in general. Its focus is on emergent business processes. It explains process architecture design and suggests design strategies for process architecture in emergent organisation. The theoretical model is based on the postulate of emergent organization, which is evidenced in the next section. Definition of business process in the context of emergent organization is then discussed. These preliminaries aside, the theoretical framework for the theoretical model is outlined in the following two sections. First, the Theory of Deferred Action, a process theory of action and design, is outlined as the base theory for developing the theoretical model. Then the 'non-trivial machine' cybernetic concept and the active model modelling concept are presented. The theoretical model draws on these concepts to improve EBP design. This sets the background for elaborating the theoretical model of process architecture design for EBP. The theoretical model is illustrated in the penultimate section with exemplar process-oriented IT systems. The concluding section is a summary and description of further ongoing theoretical and design research in process architecture design.

Emergent Organization

A business model and its internal logic can be specified to form the basis for process architecture design. Many entities composing business processes can be predicted. In a supply chain the supplier, materials, quantities, locations, and times when required can be predicted. In the main, business rules, organizational processes, procedures, and policies can be determined, predicted, and specified. Business activities engendered by implementing business rules can be used to design the activities of business processes.

However, emergent organization differs from organization that can be so specified. By definition, emergent organization cannot be determined, predicted, and specified. The term 'emergent organization' is used to describe three related process architecture design problems: emergent organization, business change and associated change management, and emergent business processes (non-standardizable processes).

Emergent organization posses a problem for process architecture design. Emergence is a characteristic of organization affecting routines, structure, and process information needs. Feldman's (2000) study of organizational routines shows how even routines are a source of continuous emergent change. A later study shows organizational structures to be emergent, affecting organizational resources planning (Feldman 2004). In organizational knowledge management research, emergence effects organizational knowledge processes (Clarke and Patel 1995; Truex D P, Baskerville R et al. 1999). Markus et al. (2002) identify organization design knowledge processes as emergent having implications for designing supporting IT systems.

Frameworks and theories are proposed to explain the effect of emergent organization on IS development (Clarke and Patel 1995; Truex D P, Baskerville R et al. 1999; Markus M L, Majchrzak A et al. 2002; Warboys B, Snowdon B et al. 2005). Patel (2006) makes emergence a central postulate for designing artefacts that are connected to human (organised) action. 'Emergence is an unpredictable affect of the interrelatedness of multifarious purposes and the means to achieve them that is characteristic of social action. By implication, emergence is the non-specifiable constraint on rational design because it cannot be determined as design objects, it is off-design.' (Patel, 2006:12). Emergence is sudden and unexpected change indicative of complexity in social systems. It requires process architecture responses in context.

Allied to emergent organization is the second problem of how expected or planned business change can be factored into process architecture design. Research in this area focuses on strategic organizational change. Such change management differs from emergence because change management is amenable to planning and can be predicted. Among those who have proposed change management models are Gordon et. al's (2000) integrated model of change forces focusing on strategic change and Boddy et al.'s (2000) model of supply chain partnering. The third problem of EBP (non-standardisable processes) is discussed in the next section.

Business Process

Business processes are of two types standardisable and non-standardisable. The attributes of standardisable processes can be known in advance making the input-process-output relationship invariant and more amenable to design. A business process has specific inputs that are converted into predetermined outputs by a series of value-added tasks for the benefit of customers and resulting in a revenue stream for the business. Such a definition of business process assumes that inputs and outputs are invariant, implying also that the process by which the inputs are converted into outputs is invariant too. Standardisable business processes contain predictable routines and structure. A production process would have certain plant and equipment, human resources, energy and material inputs that are converted into products/services and revenue. Invariant process can be well-defined processes with high volumes, low variation in order and delivery, and short lead-times, where complexity is less and experiential learning is useful.

In contrast, attributes of non-standardisable processes (emergent business processes) only become cognizant in context, making the input-process-output relationship variant and less amenable to specification and predetermined design. Non-standardisable business processes are non-routine and contain many emergent properties that cannot be known in advance. Project management business processes and jobbing process are examples of non-standardisable business processes. Routines and structure are not possible to determine for non-standardisable processes. Non-standardisable processes are highly affected by emergence. As evidenced above, emergence affects even organizational routines and structure.

Research into manufacturing processes shows the uniqueness quality of non-standardisable business processes. Hayes and Wheelwright's (1979) product-process matrix is a tool for analyzing the strategic relationship between the product life cycle and the technological life cycle. It characterises the production process as evolving and staged, moving from highly flexible, high-cost process towards increasing standardization. The stages themselves are characteristic of different process structures found in other business activity such as projects, jobbing, and one-off processes. These are non-standardisable processes whose input-process-outputs vary. Such processes are evident in organizational work involving innovation, knowledge management, project management and other knowledge-intensive business processes. Marjanovic (2005) concludes that knowledge-intensive processes cannot be 'fully pre-defined' and, for this reason, 'automation of this process is neither desirable nor possible.' Here we argue that it is possible to apply IT to non-standardisable or emergent processes.

Theoretically, the key to information management is process architecture design. Data and information is the reciprocal of business process and organisational structure. Ould (2003) classifies process into: core processes, management processes, and support processes. As a support process, sales order process generates data and information on customers, products, delivery and other support activities. Such processes are data intensive. As a management process, process management is highly information-intensive supporting management decision-making. IT is central to process architecture design because it can capture process data and process it to deliver process information for management. IT enables processes and is the bas is for designing process management systems.

The term emergent business process (EBP) describes non-standardisable processes that are dynamic, evolving, knowledge-intensive business processes (Marjanovic O 2005). Markus et al. (2002), define EBP as:

- an emergent process of 'deliberations' with no best structure or sequence;
- highly unpredictable potential users and work contexts; and
- information requirements that include general, specific and tacit knowledge distributed across experts and non-experts.

Markus et al. (2002) cite strategic business planning, new product development, and organisation design as examples of EBP. To design process architecture for emergent organization, there are four significant aspects of EBP to be considered:

Predictable Business Change

Process architecture should be designed for ease of change to facilitate predictable or planned business change. Such business change can be predicted and planned, and it requires process flexibility. Shaw et al. (2007) define flexity as 'the ability to change organizational capabilities repeatably, economically and in a timely way.' (p.92) Predictable business change occurs because organizations want to improve their performance, efficiency and effectiveness to counter competition and to respond to market change. The elements of the change programme are known. Current process management, modelling, and design techniques assume predictability.

Emergence

Emergence differs from predictable business change or planned change because it is unpredictable and sudden. Process and information requirements emerge. Emergent process has no best structure or sequence. It is not specifiable for the purpose of deliberate design. However, EBP do coexist with standardisable processes.

Process Actors

Emergence suggests a greater role for organizational actors involved in process enactment. Ould (2003) calls them 'process actors.' They are the process owners who are responsible for the performance and continuous improvement of processes. They are prerequisite of process design and Ould proposes that they should be enabled to design processes. Process actors capable of designing are necessary for EBP, which is a key concept of the Theory of Deferred Action discussed in the next section.

Reciprocity

Organizational process design and IT systems design should be consistent. Snowdon and Kawalek (2003) observe that the design of IT systems affects organization and the design of organization affects IT systems. There is a reciprocal relation between them. The emergence that affects organization also has an impact on the supporting process architecture.

These four issues are integral to process architecture design. What is an appropriate conceptual basis for process architecture design in this context? How should process support IT systems be conceptualised? A potential action and design theory capable of addressing these issues is outlined next. It forms the basis for the proposed theoretical model.

The Theory of Deferred Action

Gregor (2006) elaborates the nature of theory in information systems by demarcating five interrelated types: (1) theory for analysing, (2) theory for explaining, (3) theory for predicting, (4) theory for predicting and explaining, and (5) theory for action and design. The theoretical model presented in the next section draws on the Theory of Deferred Action, which is a 'theory for action and design.' It is a 'process theory' in terms of Sol's (1995) classification discussed in the introduction.

Rather than explain a phenomenon, action and design theories seek to develop 'usable knowledge' that can be applied to design. Walls et al. (1992) propose the IS design theory; Bell (1993) proposes a database design theory from the perspective of organization; Markus et al. (2002) propose a theory to design IT systems to support the work of organizational design; and Arnott (2006), one for the design of decision support systems. These theories aim to inform the design of IT artifacts. Shaw et al., (2007) propose the business process management system pyramid architecture to design process architecture that evolves with business change.

The Theory of Deferred Action proposes three design dimensions: planned action, emergence, and deferred action. The correlation of the planned action and emergence dimensions determines types of organization and systems design possible, as illustrated in Figure 1. The planned action dimension is typical of the ontological assumptions made in existing approaches to process architecture design. It assumes that design objects can be predetermined and specified. It results in the specified systems type indicated at point B. Existing process modelling techniques seek such specifiable design objects. Only those design objects not affected by emergence can be specified.

By introducing the emergence design dimension, the specifiability of design objects becomes less. Since the object of design will emerge in the future they cannot be specified. How a particular design will be used in emergent conditions cannot be specified. To cope with this type of design the theory proposes the notion of 'deferred action,' the third design dimension. Deferred action assumes that actual action is superior to any formal design in particular contexts and facilitates such action in the designed artefact. Catering for deferred action results in the 'deferred systems' type depicted at point A. Deferred systems are a 'a way of achieving formal objectives that combines knowable rules and procedures with actuality' and they are 'inherently future-oriented...' (Patel 2006). Mathematically, a deferred system is a continuous system with much randomness. This facilitates design objects that become necessary because of emergence.

Critically, the theory introduces the deferred action dimension as a synthesised element to design for emergent organization. The deferred action construct accounts for human action and organizational behaviour, or emergent organization, evidenced by the literature earlier. It thus synthesises planned action and deferred action necessary because of emergent organization. The deferred systems type is consequently a synthesis of planned action and action necessary because of emergent organization, deferred action. This is emergent action that cannot be predetermined as design objects.

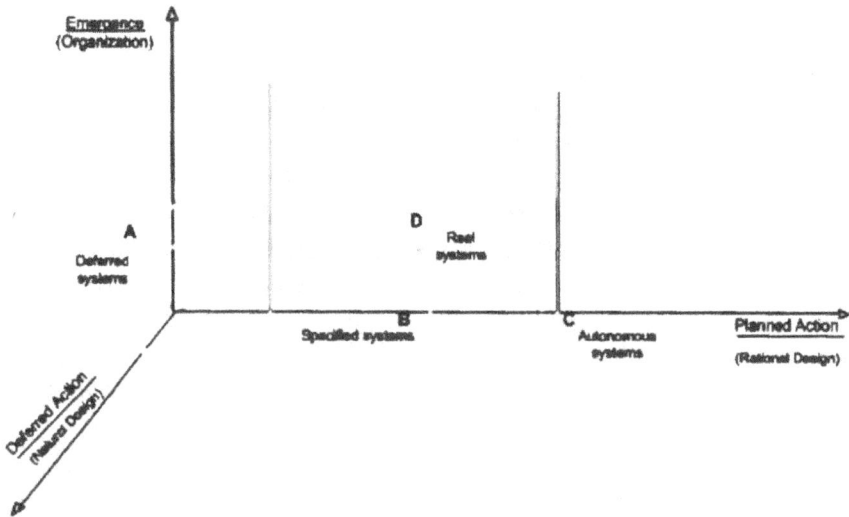

Figure 1: Types of Organization and Systems Design

The theory renders process architecture design for EBP, and emergent organization generally. This is done as the deferred systems type. EBP can be designed as deferred systems. Elliman and Eatock (2005) have applied the 'deferred design decisions' design principle, stemming from the theory, to develop IT systems to support non-standard legal arbitration processes. Sotiropoulou and Theotokis (2005) have applied the theory to develop e-government systems using service-oriented process architecture (service-oriented architecture is cited later as an example of the proposed theoretical model). The theory has been applied to develop tailorable information systems (Theotokis, Gyftodimes et al. 1996; Stamoulis D, Kanellis P et al. 2001) and e-learning systems (Dron, Boyne et al. 2003), where learning itself is characterised as a deferred system because of the temporal and cognitive distance required for learning to happen. These applications of the theory in diverse fields establish its generality.

Cybernetics and Active Models

Formal design needs to cater for variant behaviour required in actual organizational situations. The deferred action construct enables such variant behaviour in formal design. Modelling this type of organizational behaviour is discussed in this section. Deferred action can be operationalised with the cybernetic concept of 'non-trivial machine' and the 'active model' modelling type. Emergence requires designed process architecture to cater for variance, conceptualised as the non-trivial machine, and maintain a link with ongoing organisation, conceptualised as an active model.

Since deferred action cannot be predicted because of emergence, it is necessary to design systems whose input-process-output structure is variant. Foerster (2003) defines a 'non-trivial machine' as having a variant input-process-output structure. Such a system is unpredictable—the quality we seek for EBP design and emergent organization design generally, because its outputs would vary even if the inputs remain the same. The key is processes taking shape in particular situations. The non-trivial machine concept characterises well non-standardisable process whose complexity is greater—involving functional groups, strategic business units and even different companies. Such processes can be one-off and/or take long time to complete, for example in aerospace, capital goods, pharmaceutical and industrial engineering.

Modelling process architecture as an 'active model' links it well actual organization. The relevance of active models is more general. Groth (1999) conceptualises organization as active model. An active model maintains a synchronized link with the subject that it models (Snowdon B and Kawalek P 2003), in this case business processes. Warboys et al., (2000) used active models to develop process-oriented IT support systems. However, active models do not realise the non-trivial machine because the input-process-output is invariant in active models. But active models are important in the proposed theoretical model because they maintain a synchronization link with the subject—in this case EBP.

How should the design/designer be conceptualised in the context of emergence? With a variant input-process-output, the question of who designs it arises. Since emergence precludes complete predetermined designed, process actors should be enabled to design EBP. Additionally, both the non-trivial machine and active models imply an active role for process actors in designing systems. In this context, process actors are termed 'active designers.'

This has theoretical implications for process modelling techniques. There are few theoretical views on business process modelling. Melão and Pidd (2000) note a conceptual framework for process modelling, the role activity diagram approach. Melão and Pidd's own conceptual framework is a taxonomy of extant approaches to process modelling. They classify the approaches into: deterministic machines, complex dynamic systems, interacting feedback loops, and social constructs. None of this theoretical work however addresses emergent organization and EBP as characterized in this paper. The modelling techniques intrinsically result in static models, as opposed to active models type required for emergent organization. The implications of emergence for processes modelling are considered after next presenting the theoretical model for process architecture design.

Theoretical Model of Process Architecture Design for Emergent Organization

Shaw et al. (2007) define a model as a 'planned abstraction of reality represented in a form that is usable by a human.' (p.95). A theoretical model for designing is a planned abstraction based on some theory. Here the theoretical model is based on the Theory of Deferred Action. Its purpose is to support humans' design activity.

Akin to design models, a theoretical model can perform three functions: it can be explanatory, it can contain reasoning facility, and it can be basis for designing. An example of a theoretical model that explains a phenomenon is Currie and Parikh's (2006) integrative model of value creation from web services. Theoretical mathematical models are most powerful for reasoning. No reasoning models exist for process architecture design. Snowdon and Kawalek's (2003) active meta-process model is a conceptual model for designing process architecture. The theoretical model developed in this paper is for designing. It is an 'action and design' theoretical model, rather than simply an explanatory model, but does not contain reasoning power. It draws on and is deduced from the Theory of Deferred Action, the non-trivial machine concept, and active models discussed above.

The theoretical model is for improving EBP design but it can be used for standardisable business process design too, as standardisable processes are affected by emergence in the long run. It is capable of addressing predictable business change and unpredictable EBP requirements. Change management is relatively non-problematical because the associated process architecture can be predicted and specified. EBP cannot be similarly specified because they emerge in unpredictable and sudden ways. The unpredictable class of business processes cannot be pre-defined and pre-specified for design purposes. The cause of this unpredictability is emergent factors. The theoretical model helps improve our understanding of how to design EBP by understanding the effect of emergence on the design of systems in general.

Theoretical Constructs

Designing for emergent organization is problematical and complex. The design has to cope with endogenous and exogenous business factors, some predictable and others emergent. There are prominent established research streams relevant for understanding these factors, which also form the basis for proposing the theoretical model. The theoretical model contains five constructs, shown in Table 1, drawn from the information systems development (specified design), organization studies (emergent organization), deferred action, deferred design, and ebusiness model research streams.

Table 1: Construct Definition

Construct	Definition	Evidence Base
Specified design	Design that requires complete specification of requirements. Specification is central to designing process architecture or 'infrastructure software' and information systems. Design is a rational process.	(NATO 1968; Demarco 1978; Mumford, Hirschheim et al. 1985)
Emergence	Sudden and unpredictable occurrence of events that make rational design by specification alone impossible.	(Feldman, 2000; Feldman, 2004; Patel 2006)
Deferred action	Consequence of relating specified design with emergent organization is deferred action. Design that cannot be predicted because of emergence is deferred to organizational actors or 'process actors.'	(Elliman T and Eatock J 2005; Patel 2006)
Deferred design	Deferred design is design by 'action designers' (organizational actors) within formal design to cope with unknowable emergence or 'equivocal reality'.	(Purao S, Truex D et al. 2003; Dron J 2005; Elliman T and Eatock J 2005; Patel 2006)
Process architecture	Process architecture is composed of artifacts that are a combination of business process design and supporting IT systems design. The process architecture is a socio-technical system.	(Beeson I, Green S et al. 2002) Marjanovic O 2005; Snowdon B, Warboys B et al. 2006)
ebusiness model	eBusiness model is composed of business processes and supporting IT systems designed to generate and sustain revenue streams. It is based on the concept of 'business model' for producing, delivering and selling product or services of value to customers and capable of creating wealth.	(Timmers 1999; Margretta J 2002)

Specified Design

The term specified design is used to describe design that requires complete specification of requirements based on construing design as a rational process. Specification is central to rationally designing process architecture or 'infrastructure software' and information systems. Information systems development approaches (Demarco 1978; Yourdon and Constantine 1978; Gane and Sarson 1979) and IS development methodologies (Martin J and Finkelstein C 1981) make specification a prerequisite for designing rationally. The resultant systems models are static as they contain no link with the subject domain modelled.

Dearden (1972) commented early on the limitations of rational or specified design. He stated that it is impossible to pre-design all the information requirements for a company. Attempts to cope with business change within this paradigm result in system evolution (Snowdon B, Warboys B et al. 2006) as a software engineering solution to business change. Business change has engendered proposals for dynamic modelling (Giaglis 1999) and suggestion for 'postmodern software development' (Robinson, Hall et al. 1998). Swartout and Balzer (1982) sought to break form the rational design paradigm by proposing that 'requirements' and 'implementation' are not discrete but 'intertwined.'

Emergent Organization

Emergence is the efficient cause of the difficulties with specified design. Specifically, for information systems design, Truex et al. (1999) propose the explicit recognition of emergence in IS development approaches. Baskerville et al., (1992) are more radical in proposing 'amethodological' approaches, in which phased development or rational design is underplayed for a continuous development approach akin to deferred systems.

Wieck (2004), the organization theorist, argues for design by 'underspecification' as a solution to business change. The basic idea is to gather a specification that forms the 'skeleton' for the design and enabling organizational actors to fill in the 'flesh' in actual organizational contexts. The proposal is a general solution to the problem of emergent organization too. It is key to the proposed theoretical model, as it underpins the deferred action, deferred design, and process architecture constructs of the model.

Process architecture design has moved away from individual 'applications' to organisation-oriented design, centrally recognising organizational change and complexity. McDermid (1994) calls approaches to requirements engineering that focus solely on applications functionality 'orthodox.' He calls for an organizational focus to address requirements comprehensively for changing organization. This organizational focus is addressed by process-oriented IT systems development and research funded by UK government agencies (Henderson 2000).

Deferred Action

The consequence of relating specified design with emergent organization is deferred action (Patel 2006). Design that cannot be predicted because of emergent factors is deferred to organizational actors or 'process actors.' Deferred action not only explicitly recognises the limitation of specified design by adhering to the notion of underspecification but, critically, provides a way forward to design systems for emergent contexts. Since design by complete specification of requirements is precluded by emergence the design of the 'skeleton,' to use Wieck's (2004) term, should include the capability to do deferred design. The notion of tailoring information, based on deferred action, for specific contextual needs was recognised earlier by Macmillan (1997).

Deferred Design

As the design of a complete artefact is not possible, deferred design is necessary. Michl (2002) regards all design as 'redesign,' meaning that all design is incomplete. In this paper it is termed deferred design, which recognises the incompleteness of design and enabling continuous design. The need for deferred design is acknowledged by IS and software researchers (Theotokis, Gyftodimos et al. 1997; Truex D P, Baskerville R et al. 1999; (Carey J E and Carlson B A 2000). Deferred design is made possible within a formally designed framework, distinguishing it from instrumentalism.

Design for changing and emerging processes is deferrable as deferred design to process actors in actual contexts. (Dron J 2005; Elliman T and Eatock J 2005; Sotiropoulou A and Theotokis D 2005). Researchers affiliated to the International Federation of Information Processing (IFIP) recognise the importance of deferred design (Purao S, Truex D et al. 2003).

Process Architecture

Process architecture is composed of artifacts that are a combination of business process design and supporting IT systems design. An example is customised XML scripting. The process architecture is thus a socio-technical system (Mumford 2000). The technical system element of the process architecture is also called 'infrastructure software.'

Process architecture may be classified using Keen and Scott-Morton's (1978) classification. They classify decision processes into highly-structured, semi-structured and unstructured, which is useful for business process design. Process architecture that supports emergent business processes is unstructured.

Ebusiness Model

An ebusiness model is a model of the future. It is a model of something that will be realised. Such models are also termed 'to-be models,' as opposed to 'as-is models' that model the current system of interest. Weill and Vitale (2001) discuss the transition that firms need to make from business models to e-business models. The centrality of business model is recognized in the literature. To-be business models improve understanding of enterprise success and are designed to produce, deliver and sell products or services that add value for customers and create wealth (Margretta J 2002). An objectified business model improves a company's knowledge of its purpose and operations, resulting in explicit organizational knowledge and explicit business value creation knowledge.

Process architecture design is effective when based on sound business model. Business models and e-business models explain how a business organization should organise its activities to create value for customers. However, the body of literature on e-business models lacks the necessary commensurate conceptualisation of the requisite process architecture design to support processes to achieve goals. Few ebusiness models recognize emergent organization. For instance, Patel (1995) proposed emergent form of IT governance to support global ebusiness models.

The constructs detailed above are related and their interrelationships result in an ebusiness model capable of emergent behaviour. The relationships are defined in Table 2. An ebusiness model is an expression of these constructs and their interrelationships. Emergence is the independent variable that effects specified design, process architecture and ebusiness model. Specified design and deferred design co-exist in the process architecture design. Deferred action is necessary when specified design is correlated with emergent organization.

Table 2: Defining construct relationships

Construct	Relationships
Specified design	Specified design is the basis for designing an ebusiness model whose process architecture is emergent.
Emergence	Emergence is the independent variable that effects specified design, process architecture and ebusiness model.
Deferred action	Deferred action is the consequence of relating specified design to emergence. Deferred action is necessary in emergent organization.
Deferred design	Catering for emergent organization is enabled by deferred design and the deferred design decisions principle.
Process architecture	Process architecture is the enabling mechanism for emergent organization and its consequence.
ebusiness model	ebusiness model is an expression of the constructs and their interrelationships. It is designed by specified design and contains deferred design capability in order to respond to emergence. The process architecture is emergent.

The model has generic applicability. Company resources and structures are organised as business processes. Since no competitive business is free of business change, the theoretical model has generic applicability to manufacturing and service sectors. Growth and innovation are important for all businesses, and they are affected by markets and competitors' actions. Central for achieving growth and generating innovation is the design of ebusiness models and business process. Innovation in particular is subject to emergence.

The central element affected by business change and emergence is business process. The ability of businesses to meet business change and emergence depends on appropriate process architecture being in place. Both growth and innovation can be facilitated on agile process architecture.

Reliability, Validity and Propositions

The theoretical model is checked in this section for reliability and validity. Reliability and validity are of interest to improve the strength of the knowledge claim. Reliability is concerned with repeatability and corroboration. Validity is concerned with the appropriateness or meaningfulness of the knowledge claim (Rosenthal R and Rosnow R L 2008). The theoretical model can be assessed for reliability and validity by checking its internal and external consistency. Internal consistency is discussed in this section and external consistency in the next section.

The internal consistency of the theoretical model depends on the veracity of the Theory of Deferred Action on which it is based. The theory is a 'far-reaching' theory (Kaplan A 1964), as it addresses the design of socio-technical systems (Mumford and Beekman 1994; Mumford 2000). The veracity of the theory is attested to by its application. Researchers have drawn on its deferred design decisions principle (Patel, 2005) to design IT systems capable of coping with organizational change and emergence (Fitzgerald, 1999; Elliman T, 2005). It is the subject of joint research proposal with the UK Ministry of Defences' Defence Science and Technology Laboratory (dstl). The theory's construct of deferred system is proposed to be realised as an IT system using technology developed by the Informatics Process Group at Manchester University's Computer Science Department.

Is it possible to design an enterprise architecture that is able to cope with both predictable outcomes and unpredictable outcomes? This depends on the veracity of the propositions deduced from the theoretical model that inform process architecture design. The explanatory capability of the theoretical model is deepened because it combines relevant constructs from cybernetics and modelling. Consequently, the value of business propositions derived from the model is improved. Research propositions are useful because they direct further research and clarify the logic of the theoretical argument. As propositions involve concepts (Whetten D A 1989), the validity and interrelations of the concepts is clarified when stated in propositional form. Propositions 'should be limited to specifying the logically deduced implications for researchers of a theoretical argument.' (Whetten D A 1989) p.492.

Three propositions are derived from the theoretical model:

- Proposition 1. Process architecture design is effective when based on a sound e-business model.
- Proposition 2. Organizational emergence affects process architecture design.
- Proposition 3. The model of EBP (non-standardisable) process evolves.

The three propositions are logically connected. An e-business model seeks to deliver business performance based on process efficiency supported by IT systems. However, organizational emergence determines what is required from the process-oriented IT systems. Therefore, as the non-standardisable processes evolve in response to emergence, so does the supporting IT systems. These propositions are illustrated in the exemplar systems discussed next.

Process-Oriented IT Systems

Contribution to improving our understanding and capability of explaining the phenomena of interest constitutes the external validity of the theoretical model. The theoretical model can be used to explain process support IT systems design and methodologies. Conceptualisation of IT systems can be improved based on the theoretical model.

There are implications of the theoretical model for models of EBP and business process modelling techniques. How does the theoretical model inform the methods used to design process architecture? What kind of IT systems design is required to support non-standard business processes and business processes requiring change? A key question is how the theoretical model can contribute to process architecture design problem-solving. The explanatory capability of the theoretical model can be demonstrated by considering exemplar systems. The kinds of design problems it has to be capable of addressing include emergent organization, business change, and competitors' moves. In this section, the theoretical model is exemplified.

Models inform process-centered IT systems design. In turn, planning and problem-solving with models require clear design and development methodology. Shaw (2007) defines a model as: 'A model is a planned abstraction of reality represented in a form that is usable by a human.' (p.95). Slack (Slack N and Lewis M 2005) proposes a model of business process technology. Larger and Horte (T Larger and Horte 2005) have developed a classification system for process technology. The two exemplar systems discussed in this section are similarly model-based systems.

Process-oriented approach to IT systems development is adopted in the research literature. This research is at the enterprise level. The UK Government's grant-awarding research body Engineering and Physical Sciences Research Council (EPSRC) supported research into enterprise process-oriented systems architecture. It focused on how organizational change affects the development of IT systems. The result of this research is process-oriented systems architecture as reported in Henderson (2000).

In terms of cybernetics, process-oriented approach can be categorized as trivial machine and non-trivial machine, explained above. Much research can be categorised as trivial machine conceptualization of process-oriented IT systems, as reported in Ould (2003) and (Henderson, 2000). Larger and Horte (2005) provide a classification of success factors for developing process technology. Alongside this research is research that can be categorized as non-trivial machine conceptualisation of process-oriented IT systems. A particular strand of interest is on active models (Warboys B, Greenwood R M et al. 2000). Active models address the problematical issue of change in business processes and the necessary commensurate change in IT systems.

Exemplar 1 Process

ProcessWeb

The ProcessWeb is an IT system developed by the Informatics Process Group at the Computer Science Department of Manchester University, UK. Process-Web adopts a process-oriented perspective to conceptualise and design systems architecture suitable for evolution, which is necessary to cope with organizational change (and emergence). Theoretically, ProcessWeb adopts a systems approach, particularly Viable Systems Model (Beer 1979) and active model perspective on business process (Greenwood R M, Robertson I et al. 1995; Warboys B 1995; Snowdon B and Kawalek P 2003; Warboys B, Snowdon B et al. 2005).

The active model provides the synchronisation link between business process change and supporting IT systems. This achieved by maintaining a synchronisation link, through the coordination layer, with the subject of the system—in this case business processes. In active models the meta-process is the process of changing a process (Warboys B, Greenwood R M et al. 2000). An active model is contrasted with a 'passive model' which is static rather than dynamic, representing the position at the point of observation and lacking an updating mechanism (Beeson I, Green S et al. 2002). Integral to the active model of business process is the process 'coordination layer.'(Warboys B, 2000). The coordination layer enables the co-evolution of business process and IT systems.

In terms of cybernetics, ProcessWeb can be categorised as a non-trivial machine conceptualization of business processes with one critical qualification. In ProcessWeb, the input-process-output relationship is non-variant, whereas in non-trivial machines it is variant. Whilst the active model provides a link with the actual business processes modelled, the system architecture of ProcessWeb does not enable variable input-process-output.

ProcessWeb illustrates four of the five constructs, and their interrelationships, of the deferred action theoretical model. The theoretical model directs modelling attention to emergent factors. In ProcessWeb, specified design is based on the active model. As the active model keeps a synchronised link with the business domain, it accounts for emergent factors. There is no direct enablement of deferred action. However, the synchronisation link in the active model is the mechanism that enables organizational actors to keep the IT system relevant to business needs. It indirectly caters for deferred action. The conceptual model of the ProcessWeb is based on business process and so its system architecture is process-oriented. As ProcessWeb is an experimental system it has no e-business model.

The three propositions deduced from the theoretical model are accounted for in ProcessWeb. Concerning proposition 1, ProcessWeb is capable of supporting any process-based business model since its architecture is process-oriented. It seeks to be effective by catering for business processes. Proposition 2 is met because organizational change, an aspect of organizational emergence, is catered for by the synchronisation link in the active model. Since ProcessWeb is predicated on software evolution, it meets Proposition 3 of non-standardisable processes evolving with the same magnitude of IT systems evolution.

Exemplar 2 Service Oriented Architecture (SOA)

The underlying design principle of Service Oriented Architecture (SOA) is the provision of software services to business on demand. The services are loosely coupled or configured to meet specific business process needs (Jones S 2005). IT system architecture, including computer networks, is accessed without constraints to deliver required services.

SOA illustrates five of the constructs and their interrelationships of the deferred action theoretical model. Specified design is based on software (and hardware) components. A software component is a system element offering a predefined service and able to communicate with other components. Since software components are 'non-context—specific,' they enable emergence through composition to be represented in IT systems. Software componentry, indirectly, makes deferred action at the level of systems designers possible (as opposed to organizational actors). Concerning the process architecture, as components are non-context—specific, they can be mapped onto existing or newly designed business processes. The technical system architecture is dis tributed computing based on application servers. SOA, within which software componentry is embedded, is suited to support emergent e-business models . Business processes and the needs of organizational actors underpin SOA. It recognises the interconnectedness of organization, data and applications. The architecture is designed to deliver computational resources on demand as required by business users, thereby serving any e-business model.

The three propositions deduced from the theoretical model are accounted for in SOA. Concerning proposition 1, since the concept of service underpins SOA, any underlying e-business model is supported. IT systems are configured to support specific, and unique, business processes. Proposition 2 is met because organizational change (and emergence) is catered for by re-configuring services. Similarly, Proposition 3 is met as non-standardisable processes are directly catered by re-configuring services as required.

Generally, the exemplar systems are indicative of a trend towards a new conceptualization of IT systems. In terms of cybernetics, the emerging conceptualisation is tending towards IT systems as non-trivial machines whose input-process-output structure is variant. This conceptualisation mirrors the deferred action theoretical model developed above and the derived propositions. It is evident in systems like ProcessWeb and emerging technologies like software componentry, SOA, and the Semantic Web.

The theoretical model has wider implications for process management. Process management involves approaches to process improvement. It encompasses interest in business process modelling languages and theory of process management change. Process management encompasses methodologies, techniques and tools to support the re(design) of business processes. Business process design is informed by business process modelling languages. Practitioners particularly are interested in modelling languages. Kettinger et al. (1997) surveyed 25 methodologies, 72 techniques and 102 tools.

In terms of the theoretical model, the current class of modelling languages focus only on explicit knowledge of the organization and the things that can be specified.

Of the four perspectives on business processes elaborated by Melão and Pidd (2000), business processes as interacting feedback loops and business process as social contracts reveal that the possibility of the exact specification of business processes is limited. Business processes as deterministic machines and business process as complex dynamic systems, the other two perspectives, require exact specification of processes. But the perspective and modelling languages used to model processes from these perspectives assume objectivity and fixed ontology.

Objectivity is not assumed in the social contract perspective. In terms of cybernetics, they all assume the possibility of business processes as trivial machine. Given emergence, it is necessary to acknowledge emergent ontology. For instance, when business partnerships form a new business vocabulary and artifacts are also likely.

Managerial Implications

The theoretical model has implications for business managers and IT managers. Managers need to reconsider the extent to which IT architecture (process architecture) and methods for developing it can be specified. The distinction between a definitive and static business and similar IT architecture becomes blurred in emergent organization. Companies procuring IT solutions from vendors who supply fixed architecture need to reconsider their IT strategy.

Management have to think about IT centrally in emergent organization. IT cannot simply be supporting function. Managers have to revise their concept of managing in an emergent organization. Since fixed ontology becomes inappropriate in emergent contexts, management constructs need to be revised to cope with emergence.

Management have to think of ebusiness models as composing two elements, the specified element and the deferred element. The specified element should be based on sound determinable business strategy and objectives. The deferred element should operate within the boundaries of the specified element. However, it is conceivable for the ebusiness model to change marginally or radically to generate new revenue streams. Existing on-line businesses have added new revenue streams by providing services or selling products that were not in the original business model. The ebusiness model needs to be specified such that its evolution is deferred.

The implication of emergence for enterprise resource planning is that such planning activities have to be redefined as continuous. Manager's ideas of plans per se need to be changed. Rather than a discrete event planning needs to be continuous activity. Resource allocations and work design would change to meet emerging market conditions and competitors actions. There is also an implication for integrated systems. Such change would also affect cost accounting, which needs to be reported to provide integrated strategic knowledge.

Conclusion

The flexibility of process architecture is important for emergent organization. A theoretical model based on the Theory of Deferred Action was elaborated. The purpose of the model is to understand better how organizational change and emergence can be catered for in IT systems supporting business processes. The model's five constructs were detailed and their interrelationships explained. Three propositions were derived from the model. The theoretical model and the propositions were exemplified in two IT systems.

The proposed model has implications for models of business processes, particularly non-standardisable or emergent business processes. It also has implications for business process modelling methodologies and techniques. The latter in turn has implications for practice, which are beyond the scope of this paper's consideration but nonetheless important.

The theoretical model has business implications that require further research. The model's deficiencies include consideration of market leadership, strategic differentiation, and revenue generation, as a minimum basis for designing business processes and process architecture. Further empirical research, particularly cases studies of IT systems purporting to cater for organizational change and emergence, are needed. The Theory of Deferred Action is the subject of proposed research collaboration with the UK government's Defence Sciences and Technology Laboratory and the Informatics Process Group (IPG) at Manchester University. The IPG collaboration will seek to technologically realise some of the theoretical constructs presented in the theoretical model.

The theoretical model suggests a research agenda directed to improving process architecture flexibility. The paper has applied the Theory of Deferred Action to business process flexibility and improved our understanding of the robustness of process architecture and limitations of specification-based design. The implications of the theoretical analysis with regard to process architecture design require further research. The important issues that have been identified theoretically as promising areas of further research include: scope of process specification, relationship between EBP and deferred action, and development of techniques to model EBP. An important question is raised. What is the right magnitude of deferred action for particular EBP? This question is synergistically related to the impact that emergence has on organizational design. Levels of emergence determine levels of deferred action required. This relationship is a central focus of further deep research at the Brunel Organization and Systems Design Center [BOSdc].

In particular, further research will focus on the problem of demarcating specifiable and deferrable business objects in ebusiness systems design. This requires clear definition of specification and deferment in terms of the business services enabled by IT systems. Understanding ebusiness modelling and ebusiness systems designing as an emergent or continuous activity is an allied further research theme. Understanding the distinction between specifiable objects and deferrable objects and emergent design can be improved as services science. A service is an interaction between a provider and a client that produces and captures value for the client. Developments in Service Oriented Architecture and web services are important but they are predicated on specified design. We seek to understand the scope of deferrable design within service science.

References

Arnott, D. (2006). Cognitive Biases and Decision Support Systems Development: A Design Science Approach. Information Systems Journal 16: 55–78.

Baskerville, R, J. Travis, Treux, D. (1992). Systems Without Method. From Proceedings: IFIP Transactions on The Impact of Computer Supported Technologies on Information System Development. 241–270.

Beer, S. (1979). The Heart of Enterprise. New York:John Wiley & Sons Ltd. Beeson, I. Green S, Sa, J. (2002). Linking Business Process and Information Systems Provision in a Dynamic Environment. Information Systems Frontiers 4 (3): 317–329. Bell, D. A. (1993). From Data properties to Evidence. IEEE Transactions on Software Engineering 5 (6): 965–969.

Boddy, D. Macbeth, D. Wagner, B. (2000) Implementing Collaboration between Organizations: An Empirical Study of Supply Chain Partnering. Journal of Management Studies 37 (7): 1003–1017 Nov 2000.

Carey, J. E. and Carlson, B. A. (2000). Deferring Design Decisions in an Application Framework. ACM Computing Survey 32 (1).

Clarke, S. and Patel, N. V. (1995). Structure and Culture of Higher Education in Institutions—Its Impact on Information Systems Strategic Planning. Proceedings from: The Fourth International Conference of the United Kingdom Systems Society on Critical Issues in Systems Theory and Practice. Plenum, London.

Currie, W. L. and Parikh, M. A. (2006). Value Creation in Web Services: An Integrative Model. Strategic Information Systems 15: 153–174.

Dearden, J. (1972). MIS is a Mirage. Harvard Business Review 50 (90–99).

Demarco, T. (1978). Structured Analysis and System Specification. New York: Yourdon.

Dron, J. (2005). Epimethean Information Systemsd: Harnessing the Power of Collective in e-Learning. International Journal of Information Technology Management. 4 (4): 392–404.

Dron, J. Boyne, C. Mitchell, R. (2003). Evolving Learning in the Stuff Swamp. In Patel N. V. (Ed.) Adaptive Evolutionary Information Systems. 211–228. London: Idea Group Publishing.

Elizinga, D. Horak, T. Chung-Yee, L Bruner, C. (1995). Business Process Management: Survey and Methodology. IEEE Transactions on Engineering Management 42: 119–128.

Elliman, T. and Eatock, J. (2005). Online Support for Arbitration: Designing Software for a Flexible Business Process. Int. J. of Information Technology and Management 4 (4): 443–460.

Feldman, M. S. (2000). Organizational Routines as a Source of Continuous Change. Organization Science 11(6): 611–629.

Feldman, M. S. (2004). Resources in Emerging Structures and Processes of Change. Organization Science 15(3): 295–309.

Fitzgerald, G., Philippides, A. (1999). "Information Systems Development, Maintenance and Enhancement:Findings from a UK Study." International Journal of Information Management 19: 319–328.

Foerster, H. V. (2003). Understanding Understanding. New York: Springer-Verlag.

Gane, C. and Sarson, T. (1979). Structured Systems Analysis: Tools and Techniques. New York: Prentice-Hall.

Georgakopoulos, D. Hornick M. Sheth, (1995). An Overview of Workflow Management: From Process Modelling to Workflow Automation Infrastructure. Distributed and Parallel Databases 3: 119–153.

Giaglis, G. M. (1999). Dynamic Process Modelling for Business Engineering and Information Systems Evaluation. Department of Information Systems and Computing. London, Brunel University.

Gordon, S. S. Stewart, W. H. and Sweo, R. (2000) Convergence versus Strategic Reorientation: The Antecedents of Fast-Paced Organizational Change. Journal of Management 26 (5). 911–945.

Greenwood, R. M., Robertson, I. Snowdon R. A., Warboys, B. (1995). Active Models in Business. Processing from: The annual conference on business information technology.

Gregor, S. (2006). The Nature of Theory in Information Systems. MIS Quarterly 30(3): 611–642.

Groth, L. (1999). Future Organisation Design. Chichester:Wiley.

Hayes, R. H. and Wheelwright, S. C. (1979). Link Manufacturing Process and Product Life Cycles. Harvard Business Review January.

Henderson, P. (2000). Systems Engineering for Business Process Change. London: Springer.

Jones, S. (2005). Toward an Acceptable Definition of Service. IEEE Software 22(3): 87–93.

Jones, S. (2005). Toward an Acceptable Definition of Service. IEEE Software 22 (3): 87–93.

Kaplan, A. (1964). The Conduct of Inquiry: Methodology for Behavioural Science. Scranton PA: Chandler.

Keen, P.G. and Scott-Morton, M.S.S. (1978). Decision Support Systems: An Organizational Perspective. Reading: MA:PAddision-Wesley.

Kettinger, W. and Grover, V. (1995). Toward a Theory of Business Process Change Management. Journal of Management Information Systems 12: 9–30.

Kilduff, M. (2007). Celebrating Thirty Years of Theory Publishing in AMR: Award-Winning Articles from the First Two Decades Revisited. The Academy of Management Review (AMR) 32 (2): 332–333.

Larger, T. and Horte, S-A. (2005). Success Factors for the Development of Process Technology in Process Industry Part 1: A Classification System for Success Factors and a Rating of Success Factors on a Tactical Level. International Journal of Process Management and Benchmarking. 1(1): 82–103.

Macmillan, H. (1997). Information Systems: Four Good Questions for the Board. In Macmillan, H. and Christopher, M. (eds.) Strategic Issues in the Life Assurance Industry. Oxford: Butterworth Hienemann.

Malao, N. and Pidd, M. (2000). A Conceptual Framework for Understanding Business Processes and Business Process Modelling. Information Systems Journal 10: 105–129.

Margretta, J. (2002). Why Business Models Matter. Harvard Business Review 80(5): 86–93.

Marjanovic, O. (2005). Towards IS Supported Coordination in Emergent Business Processes. Business Process Management Journal 11(5): 476–487.

Markus, M. L. Majchrzak, A. Gasser, L. (2002). A Design Theory for Systems that Support Emergent Knowledge Processes. MIS Quarterly 26: 179–21.

Martin, J. and Finkelstein, C. (1981). Information engineering. Englewood Cliffs, Prentice Hall.

McDermid, J.A. (1994). Requirements Analysis: Orthodoxy, Fundamentalism and Hersey. In Jirotka M and Goguen J. (eds.) Requirements Engineering: Social and Technical Issues. London: Academic Press.

Michl, J. (2002). On Seeing Design as Redesign. Scandinavian Journal of Design History. 12: 7–23.

Miers, D. (1994). Use of Tools and Technology Within a BPR Initiative. In Coulson-Thomas C. (eds.) Business Process Re-Engineering: Myth and Reality. Amsterdam, North-Holland.

Mumford, E. (2000). A Socio-Technical Approach to Systems Design. Requirements Engineering 2 (5): 125–133.

Mumford, E. and Beekman, G. J. (1994). Tools for Change & Process, A socio-technical approach to business process re-engineering. Netherlands: CSG Publications.

Mumford, R., Hirschheim, R. et al., (eds.) (1985). Research Methods in Information Systems. New York: North Holland.

NATO, S. C. (1968). Conference on Software Engineering, Germische.

Ould, M. A. (2003). Preconditions for Putting Processes Back in the Hands of their Actors. Information and Software Technology. 45: 1071–1074.

Patel, N. V. (1995). Emergent Forms of IT Governance to Support Global E-Business Models. Journal of Information Technology Theory and Application. 4 (2): 1–24.

Patel, N. V. (2005). Sustainable Systems: Strengthening Knowledge Management Systems with Deferred Action. International Journal of Information Technology Management. 4 (4): 344–365.

Patel, N. V. (2006). Organization and Systems Design: Theory of Deferred Action. Basingstoke: Palgrave Macmillan.

Purao, S. Truex, D. Cao, L. (2003). Now the twain shall meet: Combining social sciences and software engineering to support development of emergent systems. Proceeding from: Ninth Americas Conference on Information Systems, Tempa, Florida.

Reijswoud, V. Mulder, H. Dietz J. (1999). Communicative Action-Based Business Process and Information Systems Modelling with DEMO. Information Systems Journal 9: 117–138.

Robinson, H., Hall, P. Hovenden, F. Rachel J. (1998). Postmodern Software Development. The Computer Journal. 41 (6): 363–375.

Rosenthal, R. and Rosnow, R. L. (2008). Essentials of Behavioural Research. New York: McGrawHill.

Shaw, D. R. Holland, C. P. Kawalek, P. Snowdon, B. Warboys, B. Elements of a business process management system: theory and practice. Business Process Management Journal. 13 (1): 91–107.

Slack, N. and Lewis, M. (2005). Towards a Definitional Model of Business Process Technology. International Journal of Process Management and Benchmarking 1 (1): 3–24.

Snowdon, B. and Kawalek, P. (2003). Active Meta-Process Models: A Conceptual Exposition. Information and Software Technology. 45: 1021–1029.

Snowdon, B. Warboys, B. Greenwood R. M. Holland, C. P. Kawalak, P. Shaw, D. R. (2006). On the Architecture and form of Flexible Process Support. Software Process Improvement and Practice in press.

Soh, C. and Markus, M. L. (1995). How IT Creates Business Value: A Process Theory Synthesis. Proceedings From: 16th: International Conference on Information Systems, Amsterdam.

Sotiropoulou, A. and Theotokis, D. (2005). Tailoring Information Systems: An Approach Based on Services and Service Composition. Int. J. Information Technology and Management. 4 (4): 366–391.

Stamoulis, D. Kanellis, P. Martakos, D. (2001). Tailorable Information Systems: Resolving the Deadlock of Changing User Requirements. Journal of Applied System Studies 2 (2). Swartout, W. and Balzer, R. (1982). On the Intertwining of Specification and Implementation. Communications of the ACM. 25 (7): 438–440.

Theotokis, D., Gyftodimes, G. Geogiadis, P. Philokyprou, G. (1996). Atoms: A Methodology for Component Object oriented Software Development Applied in the Educational Context. Proceedings from: International Conference on Object Oriented Information Systems, London, UK, Springer.

Timmers, P. (1999). Electronic Commerce. Chichester: Wiley.

Truex, D. P. Baskerville, R. Klein, H. K. (1999). Growing Systems in Emergent Organisations. Communications of the ACM. 42 (8).

Walls, J. G. Widmeyer, G.R. El Sawy, O.A. (1992). Building an information system design theory for Vigilant EIS. Information Systems Journal. 3 (1): 36–59.

Warboys, B. (1995). The software paradigm. ICL Technical Journal 10 (1): 71–79.

Warboys, B. Greenwood, R. M. et al. (2000). Modelling the Co-Evolution of Business Process and IT Systems. In Henderson, P. Systems Engineering for Business Process Change. London, Springer.

Warboys, B. Snowdon, B.Greenwood, R. M. Seet, W. Robertson, I Morrison, R. Balasubramaniam, D. Kirby, G. Mickan, K. (2005). An Active-Architecture Approach to COTS Integration. IEEE Software. 20–27.

Weber, R. (2003). Editor's comments: Still Desperately Seeking the IT artefact. MIS Quarterly. 27(2): iii-xi.

Weick, K. E. (2004). Rethinking Organizational Design. In Boland, R. J. and Callopy, F. (eds.) Managing as designing. Stanford, California: Stanford University Press.

Weill P. and Vitale, M. R. (2001). Place to space: Migrating to eBusiness Models. Harvard Business School Press.

Whetten, D. A. (1989). What Constitutes a Theoretical Contribution? The Academy of Management Review. 14 (4): 490–495.

Yourdon, E. and Constantine, L.L. (1978). Structured Design: Fundamentals of a Discipline of Computer Program and System Design. New York: Yourdon Press.

The Survey of the Political Costs and Firm Size: Case from Iran

Reza Tehrani, Mahdi Salehi, Hashem Valipour
and Mohammad Jahandar Lashky

ABSTRACT

The political cost is one of the most important costs and payment of companies, and companies sought to reduce these costs. According to the theory that political expenditures by Watts and Zimmerman has been presented, Politicians have the power to by using the policies of the distribution of wealth again such as taxes, aid, the contributions, insurance and etc companies under the influence of. In other words, the change in currents can cash by taxes, special rules and information pertinent to the political challenges under the influence of. The aim of this research study is the relationship between the size of the company and the political costs.

To do research the historical information of all present companies in Tehran stock exchange in the period of time 2005 to 2007 has been used. In the way of analyze information after making dependent conversional normal, seeing transmittal charts and with regard to the index R2, logarithmic models have been selected for the expression of communication.

The results of different patterns show that relation between the size of company and political costs is meaningful. It means that by increasing the costs of the company as much as political as well as it will increase. On the other hand because companies' environment in political activities and many of the cost of the political support by law, so a big part of the political challenges that the companies will be imposed is unavoidable and another part of it like celebration relations and the cost for participation in elections and help and contributions have been impossible to control and it can be with proper management and precise auditing by considering the long-term works reduced them.

Keywords: Political Costs, the Size of Company, Volume of Asset, Sale

Introduction

Nowadays' accounting in developed countries has managed key role and its valuable role and capabilities for to achieve organization goals. Also in Iran by the transition stage of the industrial and move toward scientific management, competitive market and increasing privatization process important information in decision makings and turning to economic more than before. With peer role and importance of accounting system and need more economic units to it, scientific bed there need to use accounting for more necessary and this feeling of the necessity, researcher to in regard to the political challenges that in Iran is not known study. Research is one of the pillars of growth and development in different aspects of economic, social and cultural every country. Development process based on regular programming to achieve the goals of the predetermined cause the desired result. The most important part in program collection is necessity to requirement information, because the information is one of the essential resources for economics and scientific progress in each country. In preparing information should consider to preservation and improvement comparable features and to be convenient information and financial reports of economic unities. In other words, information has been main force and society former and it can search its important role in large filed of human essential activities. Accounting knowledge has the considerable usage in all part of community because of its categorical effect on right financial works process. In the recent decades, rapid progresses of economic units' activities, with information systems complications have been intensified the necessity of prepare reliable and relative financial information by economic units. One of the most important necessities of preparing and presenting financial information with mentioned features is collecting the accounting standards and observance them in action. Usually collecting these standards based on political condition, economic situation and the most important accounting theories. Considering accounting hypothesis, theories, and accounting methods accomplisher have been followed different approaches for collecting accounting theory. The result of following these approaches is creating continued crisis in accounting that the effect of political condition on the process of collecting accounting standards added in it.

One of the aims of financial statements is reflex of the results of management supervisory or their accounting in front of resources which is under of their authority. The users of financial statements usually want to evaluate supervisory function or management accounting for economic deciding. Whereas preparing financial statements is the duty of management department, it may cause to go forward to management profit because of different reasons. One of the motives of management profit is political costs. According to the size of company theory there is a probability they by growing companies their managers causing to go forward to select the accounting rules in order to relegate the reports of income of present period to the coming period. Whereas the economic, social, and cultural environment in Iran is very different in comparison with western countries, it may be different the management motives because of it too.

Research Problem

According to political costs theory by Watts and Zimmerman, politicians have the authority to use the politics of the distribution of wealth again and with mechanisms like taxes, helps, contributions, insurance and etc, affect the companies. Also one of the effective factors on managers' wealth is cash rewards. In addition changes in cash flows can affect by tax, specific regulations and related information to political costs. So managers are forced to consider cases that affect on company and control them if it is possible.

By Watts and Zimmerman research about political costs and the relation of these costs with the size of company, was confirmed the theory that larger company suffer more political costs than smaller company. In fact they express that the amount of political costs strongly depends on the size of company (Millen, 2001, p.15). The question that is considered in this research is: "Is there any meaningful relationship between the size of accepted companies in Tehran stock market and their political costs or not?"

Most of Iranian companies' managers do not know about the costs during the activity in effective political environment on companies and suffering political costs from definition and classifying. This research is about to be or not to be the political costs and its relation with the size of new research company and aware all managers and active members of stock market and accounting and auditing specialists.

Review of Literature

Political costs theory introduced by "Watts and Zimmerman" in 1987 for the first time and the result of above research was expressive that the larger companies (Firouzi 1998). According to basis Watts and Zimmerman research several tentative researches have done that were following the testimonials to proof political costs theory. Some of the tentative researches have shown the relation between social divulgences and the size of company and also the relation between social divulgences and the type of industry, Watts and Zimmerman's answer to this question that why companies' measure to social divulgences is "They are beneficiary."

In Watts and Zimmerman essay about positive accounting theory and political costs it is said that positive accounting theory helps us to have better understanding of different accounting standards effects on different groups and why different groups want to spend their resources under the effect of determining standards process. In these articles the tax, insurance, managerial munificence costs, gratuitous helps, sport helps, export custom laws, establishing the seminars costs are mentioned as political costs.

Because of changes in cash flows new price of stock also can be affected by tax, specific regulations and the information about political costs. Managers are forced to consider cases that may effect on company. Managers have the higher motive to select the group of accounting standards which report lower income by tax, political consideration and legal cramp than selecting accounting standards which report more income.

According to Watts and Zimmerman ideas politicians have the authority to affect them by distribution of wealth again method of companies by their tax, regulations and gratuitous helps, contributions and etc. also some of the people who has authority proceed to pervade because of nationalists or limit a company, or in turn cause to create stimulates for politicians that suggest these actions in order to solve this pressure of politicians and other weighty groups. Watts and Zimmerman suggest that "companies want help from plans and tools like social responsibility advertisement on media, pervade on government and selecting the accounting rules" in order to report the minimum value of company's profit and income. By reporting lower incomes and profits, management can decrease the probability of different activities and reduce its expected costs (Millen-2001).

Belkouei and Karpics (1973) conducted a survey on 23 large American organizations which 7 of these 23 organizations were large oil companies. In all these companies there were worries about high profits. Also Watts and Zimmerman mentioned in their research that "large companies would like to commute into oil companies." Beside the company's size the type of the industry can affect political importance strongly. The situation of the companies is important because the companies which are in an industry usually have the same size, so membership in an industry is very important factor. Finding the statistical relation between the sizes of the company and selecting the accounting method can be the result of the company's size. However can say that membership a company in the special industry increase its political importance, but can not know determinant factor company's political situation only the type of industry. There are special industries that include the companies which gain high profit and general and political concentration themselves, like oil and gas companies or they are special industries that without considering to their profit people consider and political concentrations such as chemical companies.

The companies' tax is one part of the political costs. The larger companies have more effective rate of tax than smaller. According to Watts and Zimmerman's tentative research, larger companies like oil companies have higher effective rate of tax. Effective rate of tax in this review calculated according to scale paid tax on income before tax and cash turnover. Companies' rate of tax is a little paragon of their political costs because political process includes variables such as non exclusive, governmental treaties, laws, tariffs and export share. Each one of these factors affects political process and may increase or decrease the company's political costs.

So the relation between the size of the company and political costs means that large companies suffer more political costs. This result is just consideration to payments as political costs and does not affect received profits by large company in this relationship. It means that in comparison of political costs just is considered paid costs as gross and received profits don't detract of total political costs.

Leftwich and Holthausen (1983) said that the theory of the relation between the company's size and selecting accounting rule is derivative from political process economic theory. This theory used in Zimmerman's researches too, but get testimonials in Zimmerman's researches decrease the probability that the relation between the size of the company and selecting accounting rule is caused difference between the type of industry and used accounting rule. It means the difference caused accounting rules is not caused commitments changes of the type of industry. This difference does by changes of the company's size and aware selecting accounting rule.

Belkouei and Karpic (1989) recognize this subject that without quest for spending considerable social costs and decreasing the value of present profit creating phase and general attracts attention can be the justifier social costs, performance and divulgence.

Firouzi (1998) have done the research about political costs. The results are expressive that there is a close relation between the company's size at sale and political costs in each year during 1995 to 1997 and totally (the average of these three years) there was the close correlation relation while there is not this correlation between the size of company at assets and political costs in each year.

Blacconiere and Patten (1989) accepted this approach because they survey the chemical industry divulgences. Unlike Belkouei and Karpic, they don't limit their discussion to high profit and the power of exclusive rather they consider to legal costs, the costs which suffer to the companies from law and regulations. They allude to this point that can expect in chemical industry as a result of pain which enter to environment, like India tragedy, political costs increase. Also they offer to investors that increasing political costs can expect as outbreak and its relations. One of the way of decreasing political costs is decreasing divulgence at the time of engender tragedies.

Kahan (1992) reached to a latent legal snag in the survey of income management voluntary methods except divulgence about chemical companies, which was discussable in two ways:

General thoughts have mentioned to this point that the chemical industries' profit is high unconscionable. This subject has adjustment with political movement which aimed a lot of high income industries because these industries have the ability to pay suffered political costs on them. One of the ways of decreasing this pressure on chemical companies will be decreasing reported profit.

This recent approach is caused to say Kahan (1992) that this theory which the chemical companies that have high divulgence costs and have done huge investment strongly in recognizing the incomes will use the method which cause to recognizing lower profit. Pay- attention that it doesn't take a lot of time that usage of the method which cause to recognizing low incomes cause outbreak dangerous and important problems such as decreasing the stockholders' penchant to this industries.

In Godfrey and Jones (1999) studying called "the effects of political costs on grade profit" is used as the base of market's share. (Considering suspicioning Godfrey and Jones about the size of company and political costs, primary two factors –the number of members of labor union employees and manager gratuities was posed in this research too). Above studying showed that high politicalcosts are those which are under the authority of governmental laws or they are under the general strong scrutiny. Godfrey and Jones studying show that the companies which act in the banking department financial, research, fundamental, urban services suffer more political costs.

Digan and Hallam (1991) during their studying notify that the companies which have more market's share related to their industry strongly have more political costs and general scrutiny. In this studying market's share calculated according to the total of all assets.

Zimmerman has surveyed the relation between rate of tax and company's income and had studied its relation with the size of company according to sale and reached to this result that the relation between company's size and the rate of tax is not the same in all industries during the time.

Digan and Gerden (1996) found that the powerful tentative relation between environmental sensitivity of a company shows that it is done by the members of an environmental pressure groups and positive environmental divulgence level of company. Company's size has more relation with divulgence level and the companies that are in the most sensitive industries do it more. Nonetheless, Digan and Gerden (1994) tells this subject with more completely from Torthman (1979) that have testimonials for changing this theory which have been used social divulgence for giving legitimating to companies' operations which are working the industries that are disposable to threat from existence environment proponent groups. When a industry hurt the bio environment larger companies will pay more bio costs, unless they could bring testimonials to censure this theorem. Of course it is not difficult to supposition finding of the size, industry and divulgence too, which in the other studying is used in order to explain such testimonials. In fact, considering existing finding of environmental divulgence and related activities, this probability in oil companies is more than other companies. Lemman and Meyerza (1991) discuss that founded the testimonials for Zimmerman and Watts' accounting positive theory. In the same way this is exact subject whom Lemman and Kahan (1997) specified for a social positive theory and explanation environmental divulgence. Lemman and Kahan's activity (1997) need the copiously survey because it shows the type of understanding a lot of problem which discussed before. In a lot of aspects can not shirk from it yet. In the best condition, Lemman and Kahan (1997) introduced a reason for existence of different views of companies' social divulgence which it is researchers couldn't be successful in providing the tests that give theme possibility to evaluation validity and justifiability different theories. But their main request is pitch a political costs system for social divulgence. For doing it they are following to the survey environmental divulgence before and after introducing a specific environmental law. Also they mention to related losing of other tentative results and like before mention to this subject that Patten's legitimating theory (1991) caused to complete adjustment between political costs theories. Patten's discussion and related results are the same for both Digan and Gerden (1996) but Lemman and Kahan (1997) believe that these observations are more right. The related finding about topics industry's size and divulgence don't provide an enough base in this field which can have severance between these two situations (especially if accounting positive theorists don't emphasis on industry's size scales which is with high profit and exclusive behaviors) and have more relation their divulgence scales with social costs.

Unfortunately, Lemman and Kahan (1997) can't introduce the suitable test in this field because their research's discussion plan lost its relation with Watts and Zimmerman (1978) and Belkouei and Karpic (1989) theory. In this process they face with social theories that are rejected easily. Lemman and Kahan (1997) mention to this subject which most of relative studies to more political costs emphasis on this subject that the companies decreased their reported income and exactly mention to this subject that Belkouei and Karpic (1989) discussion have more relation with the beneficiary groups which blame of them because of reported profits and the companies use social costs in order to decrease the profits.

Patten (1991) says that strongly include legimitation theory as a base for social treaties subject. Considering to Setty's researches (1974) each social institute (including commercial institutes) work in the society which limited by a social treaty from that society. Therefore commercial institutes can use social divulgence for affecting on general politics. These divulgences can detect political subject or will be use to create a concept of company's social responsibility. The purpose is that put on social and political environment which has been mentioned in commercial divulgences book as Meels (1987) said.

This is attractive that John Tajer and koontz's research (1999) about Australian BHP divulgence in mine tragedy period in Papouya in New Gineh is include huge and developed discussion about political costs theory which introduce discussion about the legitimating of this approach. In fact they say that: "legitimating theory about relative motives distribution to environmental divulgence includes outspread a political frame" because social legitimating theory depends on political and economic factors, rather on the social factors. Of course there is no way to establishing relation between legal behaviors and maximizing profit motivation by managers because this behavior is introduced as a central factor of political costs offer. Nonetheless can introduce developed discussion of political costs that suggest it Panchapaksan and McKinnen (1992), Diggan and Karol (1993) and Lemman and Kahan (1997). Because it can create problems for accounting positive theories that are following to make difference themselves from other social theory in divulgencene behavior field.

Nonetheless Lemman and Kahan studies stay only as a method. Before in this discussion was mentioned to this subject that the Watts and Zimmerman's primary study about predicating on selecting accounting method and pressure groups behavior was for shirk that successful quest of American electricity company for clean weather legal standards has done by its newspaper's fan.

Kahan and others (1979) survey the United States chemical companies' behavior which was unreliable the result legal conditions too and could create on situation which the companies had a correlated operation therein chemical find the motives to affect on political activities' result rather political costs define as developed or limited. In other cases such as legitimating theory or be legal the purpose of companies reaction is decreasing or shirk from political consideration and situation of regulations and relative costs. Considering to this factor person may expect that Lemman and Kahan (1997) focused on companies' behavior more than approve the management law.

Panchapaksan and McKinnen (1992) considered seven latent probabilities for clearing social, in addition company's size discussed about high profits and created exclusives too and introduced huge social explanations considering this specific features. Market's share and investment focus could have close relation with profit value. During discussion about a industry, the number of staffs, the number of stockholders, social responsibility divulgence its under covering level can introduce social pressure discussions which some of them will have more relation with increase costs and decreasing profit and some of them don't have any relation.

In the recent research by Paksan and Australian Mckennin (1992) used another scale for testing size validity. These scales are: "market's share (sale proportions of company to sale proportions of related industry), industry which is member, return on investment (proportion of gross fixed asset to sale), the number of staffs, the number of stockholders, reporting social responsibility (Yearly divulgence level about social responsibility units) and some of news cover (the number of articles and presented news about company in tracts in a year)."

The analysis of these factors showed that the company's size, market's share, the number of staffs, the number of stockholders, social reporting and news cover have the same effect on political costs. Return on investment doesn't put on group because researches shoe that it may suffer a company more political costs according to return and low return, too.

Darough et al. (1998) was about the negative relation between the numbers of company's staffs as a political pressure index and profit retouch. The results show that the companies which have more staffs suffer more political pressure. As a result of management this type of units proceeds to decreasing profit in order to decreasing entered pressure.

Pancha, Paksan and Mckennin (1992) introduce discussion about social responsibility divulgence and have considered the society for social cost which has the same relation with that was introduced before by Belkouei and Karpic (1989). Nonetheless this subject is specific that they expound political costs as completely separated with Watts and Zimmerman (1978) idea. While the result of these explanations don't have any difference with others (like the companies which are proceeded to divulgence and companies selected as main or primary legal regulations object.

Pourheydari and Hemmati (2004) have done research about the effective factors on management which in this research had surveyed the effect of liability treaties, political costs, remuneration plans and ownership in profit retouch by manager. About the political costs has been used to yearly sale variable and the number of staffs. The result of research representative that the negative and meaningful relation between companies size (all sales) and profit retouch which is unlike the finding western researches. The result is explanatory that by growing the companies in Iran the political pressure on them doesn't increase. Vice versa, the larger companies proceed to profit retouch to present better picture from companies' performance because of different reasons. One of the used variables to survey the political pressure research was the number of staffs.

About social divulgence suggest that is used advertising war on media as tools for consideration digression of company's huge profit which these advertisement and social divulgence decrease image of using the exclusive authority more. It means that it causes showing legal companies' huge profits. However, advertising war problem on media may be accompanied a few political agreement and jobbery of exclusive.

Moses believes that whatever the companies' size is larger manager will have motive and more powerful want in order to retouch the profit. According to his belief by growing companies their answering responsibility get more too and managers put on disposable to answering to huge groups of claimants.

The Purposes of Research

Political groups suffer the costs to be effective on political process which can include the election campaign costs and sponsoring the politicians that usually companies suffer a part of these costs in power suitable. Also getting known to the effect of process and political discussion on companies cash flows have special necessity which is one of the essential and important sources of a unit. Increasing the product price and companies services is one of the factors that decrease companies' competition power. If it can recognize and control these costs, it can be useful in increasing companies' competition power in internal markets and even international markets. One of the obvious features of each research activity is opening new windows to readers and introduces them with news.

To be new the political costs conceptual in Iran and more getting know with this type of costs by companies' superior managers and active people in stock market and professional can be the originator feature researches about it. Therefore the purposes of this research in primary and secondary purposes frame are:

Primary Purposes

The survey the dimension of political costs in companies (large and small) and how political costs take effect on competition power and companies process in the future and providing existence meaningful relation between company's size and political costs, presentation a pattern for this relation.

Secondary Purposes

1. Get know and developing conceptual and the types of political costs.
2. recognizing the types of controllable and survey of solutions increasing competition power according to the type of related costs classifying.

Research Hypotheses

Companies' manager would like to select the group of accounting standards that report low income. According to it "Watts and Zimmerman" presented three main hypotheses:

1. Gratuity plan hypothesis
2. Liability to stockholder's equity hypothesis
3. Size hypothesis

According to size hypothesis whatever companies are larger its manager have more want to relegate to coming periods. According to their opinion politicians have the authority to affect companies with using wealth distribution methods again like company's taxes, rules and regulations and gratuitous helps. By watts and Zimmerman's research about the relation between company's size and political costs this hypothesis accepted that larger companies suffer more costs than smaller. This research is following to the relation between political costs and company's size in Iranian companies (Listed companies in Tehran stock Exchange as statistical society) and uses base assets volume and the amount of net sale as Scale Company's size. This research includes these two hypotheses:

1. There is a meaningful relation between political costs and company's assets volume.

2. There is a meaningful relation between political costs and the amount of company's net sale.

The Type of Research's Method

This research is following to survey the amount of correlation between main variables in small and large companies; in fact the research's method is correlation.

Information Assemblage Method

Information which is needed for calculating variables of this research is assemblage by magazines and Tehran stock market's publishing organization's yearbook, the reports of inestimable and common general societies of accepted companies in Tehran stock market and by the companies which provided financial information of the accepted companies in stock market as the software packages.

Research's literature which is the subject of second section of this dissertation is assemblage by library studies. Books, dissertation, internal and foreign magazines which are related to the research's subject are used in writing research literature.

Research Methodology

Each research should be in an owned domain in order to pre-dominate researcher on subject during all phases.

A) Research's subjective scope includes the survey of relationship between company's size and their political costs.

B) Location scope of this research includes the accepted companies in Tehran stock market.

C) Time scope of this research includes the years 2005 to 2007.

Research's Variables

Independent variable is company's size in this research which determine according to the amount of assets and companies sale. So arrange the companies according to assets volume and after that according to the amount of sale and after that by using median index divide the companies into small and large.

Dependent variable is political costs which are measured and observed by researches in order to know the effect of independent variable on it. Some of the things that were in most of companies in statistical society are:

1. Tax (which is one of the basic and important things in all selected companies as a statistical society and recognize as the compulsory political costs.

2. Insurance right (these political costs recognize as the compulsory and important things in political costs group).

3. Help to sport (which is voluntary political costs).

4. Help to educational area development (which is voluntary political costs).

5. Establishing seminars and conferences costs (which are voluntary political costs).

6. Protecting of bioenvironmental costs (which is compulsory political costs).

7. One in thousands and two in thousands costs (which is compulsory political costs).

8. Export custom laws (which are compulsory political costs).

Titled political costs things are very common things in surveyed companies. Other political costs can classify as the other political costs because of unimportant price and infrequency them in all companies (There were these political costs in some of the surveyed companies).

The relation between company's size and political costs is:

Political costs = f (company's size)

Assemblage Information Method

To assemblage information about political costs and companies' assets volume is referred to the place of keeping financial statements in stock market and needed information introduced by surveying documents and financial statements and in some cases by using compact disks (which includes the information about accepted company's financial statements in stock market) or get from stock market's web site.

Data Analyses Method

After statistic precise and extracted information (one time according to the amount of companies' assets and political costs and another time according to the amount of sale and political costs) about large and small companies primarily is following to the meaningful relationship between dependent and independent variable that is reviewable its justification correlation test path. After relation's justification have been following to introduce a suitable relation pattern that for this subject considering to transmittal diagrams and models nature has been used panel and regression methods.

Statistic Function

To survey the meaningful coefficient of correlation has been used T Test (using the T Test and comparison it with critical value in t-student table) after the relation's justifications introduce regression model that primarily spend to survey political costs (dependent variable) normal by Kolmogorov-Smirnov Test.

Don't be normal dependent variable cause to use suitable changes to normalize it. By going on defray to survey suitable patterns by using transmittal diagrams, these diagrams show the primary idea about the type of relation how, also can recognize data outlier if exist or not. The important pattern such as line, logarithmic, second grade and third ... tested to survey relation that seems to logarithmic patterns is more suitable than other models to say relation. According to models nature has been used panel analysis to estimate model and finally has use Fisher (F) Test and statistic for survey meaningful models.

Results of the Study

In this section primarily defray to description reviewed variables description statistics such as mean, median, variance and other description statistics index, usually clear variables distribution. After that has reviewed one of the most important regression analysis pre hypothesis means normal dependent variable (political costs). Don't be normal dependent variable cause to use suitable changes to normalize it. By going on defray to survey suitable patterns by using transmittal diagrams, these diagrams show the primary idea about the type of relation how, also can recognize data outlier if exist or not. The important pattern such as line, logarithmic, second grade and third ... tested to survey relation that seems to logarithmic patterns is more suitable than other models to say relation. According to models nature has been used panel analysis to estimate model. Also to survey the companies' size has been used regression analysis and divining analysis methods in large and small companies.

Table 1 Descriptive Analyses

	Valid	Mean	Median	Variance	Skewness	Kurtosis	Minimum	Maximum
Net Sale	826	4.1E+011	1.5E+011	9.1E+023	6.437	52.270	2E+008	1E+013
Total assets log	824	26.1610	26.0918	1.918	.333	.693	22.07	32.02
Total political costs log	818	22.4846	22.4897	1.639	-.067	1.061	17.17	26.55

Description Analyses

In Table 1 description statistics calculated for reviewed variable. Mean and median calculated as the most important central indexes. Comparative conformity of these two indexes in changed variable is representative complete symmetry of these variables.

The symmetry distribution used variable will be one of the advantages of a distribution to regression analysis. Other description statistic indexes such as variance, skewness, elongation, minimum and maximum is calculated in Table 1 for related variables:

Survey of Normality

One of the most important regression models pre hypotheses are remains. Normality of dependent can be guarantor remain normality. So it is better to test the dependent variable before normality process model (this method is more advantage than survey remain normality). Null hypothesis and alternative hypothesis writes like this to survey normality:

H_0: The distribution of data has normal distribution.

H_1: The distribution of data doesn't have normal distribution.

By using Kolomogorov-Smironov Test is defrauded to test above theory. The amount of meaningful level of this test is 0.000. Whereas this number is less than 0.05, null hypothesis is rejected in %95 confident's level then. One of the usual solutions to normalize data is using suitable changes. Logarithmic change considering data distribution (skewed to right) can be a suitable change. The amount of probability is increased to 0.21 after logarithmic change, so data are normalized. Table 2 and Graph 1 show the results of testing the hypotheses.

Table 2. One-Sample Kolmogorov-Smirnov Test

		Total political costs	Total political costs Log
N		818	818
Normal Parameters	Mean	1.4E+010	22.4846
	Std. Deviation	3.0E+010	1.28041
Most Extreme	Absolute	.322	.037
Differences	Positive	.291	.037
	Negative	-.322	-.028
Kolmogorov-Smirnov Z		9.212	1.063
Asymp. Sig. (2-tailed)		.000	.209

a. Test distribution is Normal.
b. Calculated from data.

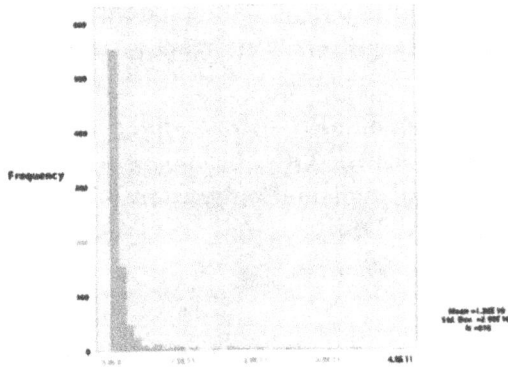

Graph 1. Total political costs

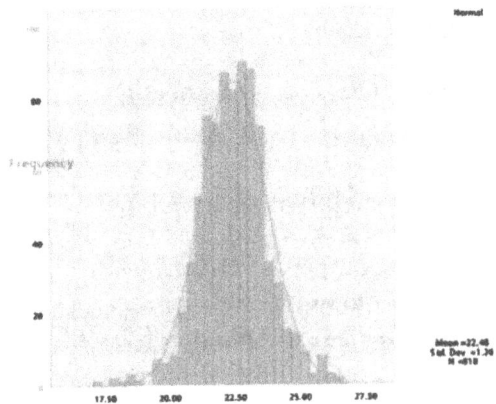

Graph 2. Total political costs log

Transmittal Diagrams

By using transmittal diagram can recognize primary pattern idea between variables. Also this diagram will show the data outlier. In bottom diagrams the logarithm of total political costs are drawn against total of assets and net sale. Line pattern as the most important pattern can't be the most suitable pattern because it doesn't show the changes of this model. The suitable pattern is recognized by using statistical indexes means the amount of R^2 and F.

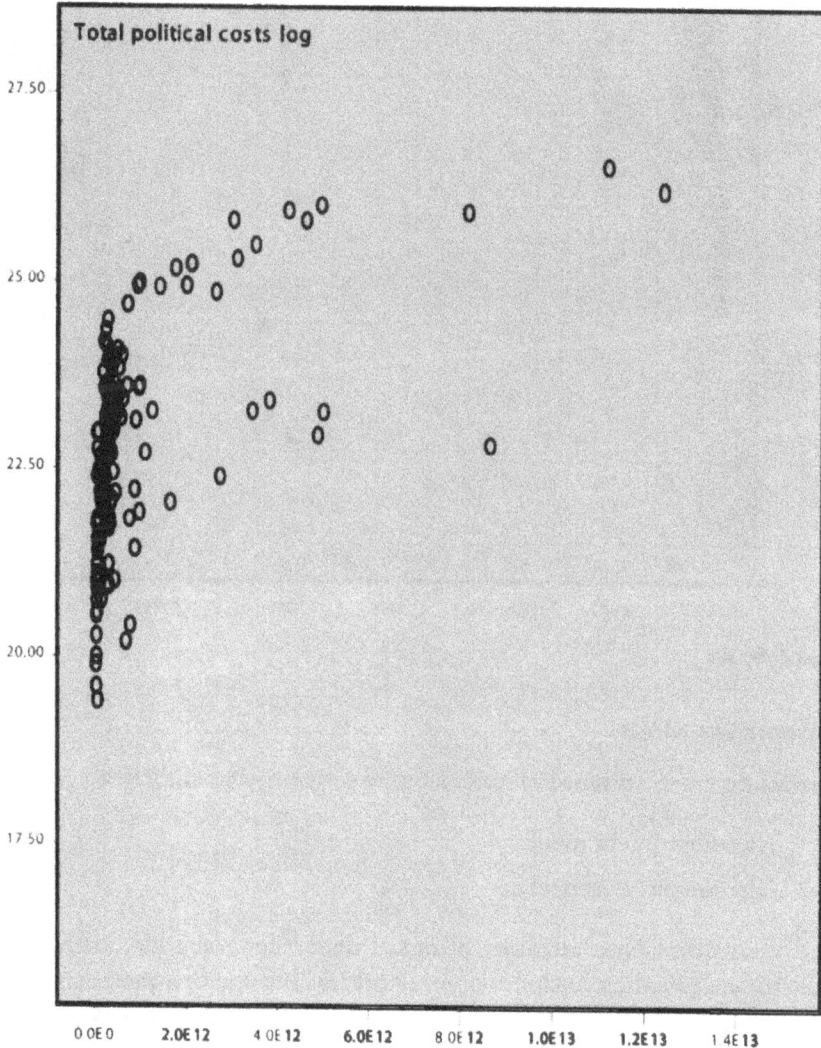

Graph 3. Total assets

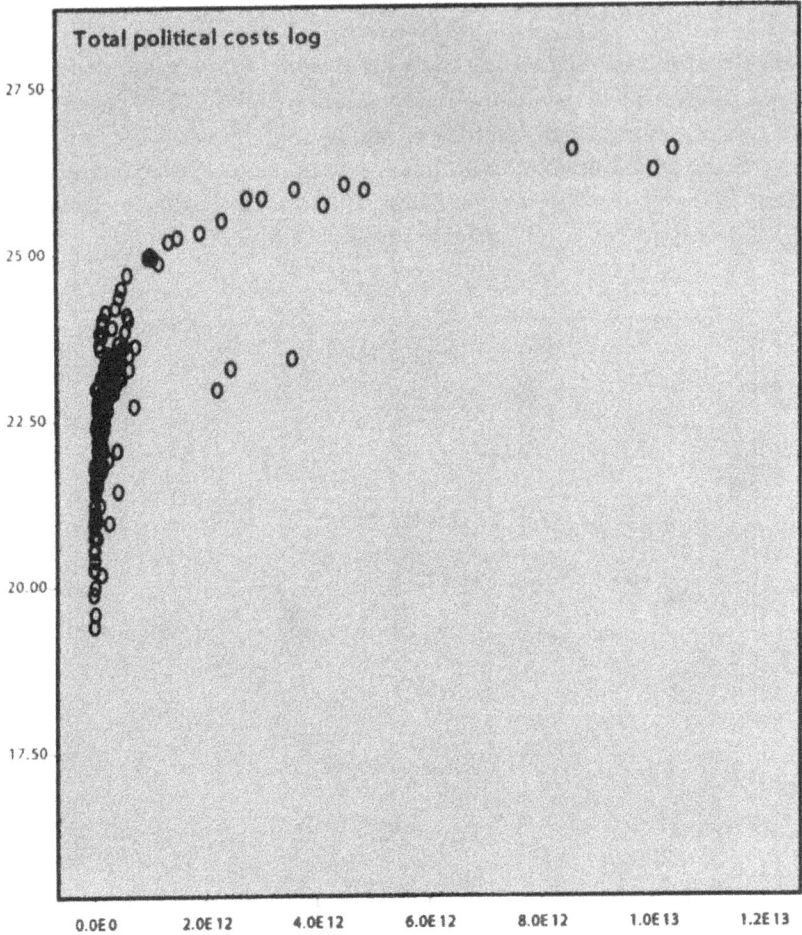

Graph 4. Net sale

Recognizing Model

In selecting model in statistical discussion two aspects are considered:

1. The attention of model
2. The simplicity of model

It means that when attention of model doesn't have sensible difference will select the model which includes more simplicity, but the first precedence of selecting model is high attention model. In regression analysis the most important index is coefficient of determination or the value of R^2 to evaluation of attention. For the total of assets variable the value of R^2 for logarithmic model is higher than other models considerably. The value of R2 is equal to 0.4 and nearest value of R2 to this model is related to third grade model which the value of it is 0.26. So the logarithmic model is suitable model undoubtedly.

Curve Fit

Table 3. Case Processing Summary

Case Processing Sumary

	N
Total Cases	1059
Excluded Cases a	247
Forecasted Cases	0
Newly Created Cases	0

a. Cases with a missing value in any variable are excluded from the analysis.

Table 4. Table Dependent Variable: Total political costs logarithm

Table Dependent Variable: Total political costs logarithm

	R Square	F	DF1	DF2	Sig.	Constant	b1	b2	b3
Logarithmic	0.403	546.107	1	810	0.000	7.121	0.587		
Quadratic	0.225	117.161	2	809	0.000	22.049	9.10E-013		
S	0.235	248.720	1	810	0.000	3.125	-1E+009		
Exponential	0.109	98.931	1	810	0.000	22.248	1.26E-014		

Note: Independent variable is total assets

Also for the second model means net sale the value of R^2 of logarithmic model sensibility is higher than other models. The value of R^2 in this model is 0.64. So it is needed to use logarithmic of sale instead of sale.

Curve Fit

Table 5. Case Processing Sumary

Case Processing Sumary

	N
Total Cases	1059
Excluded Cases a	44
Forecasted Cases	0
Newly Created Cases	0

a. Cases with a missing value in any variable are excluded from the analysis.

Table 6. Table Dependent Variables: Total costs Logarithm

Table Dependent **Variables: Total costs** Logarithm

Equation	Model Summary					Parameter Estimates			
	R Square	F	DF1	DF2	Sig.	Constant	b1	b2	b3
Linear	0.233	246.682	1	813	0.000	22.205	6.40E-013		
Logarithmic	0.642	1455.139	1	813	0.000	3.209	0747		
Inverse	0.080	70.562	1	813	0.000	22.513	-2E-009		
Quadratic	0.325	195.258	2	812	0.000	21.995	1.47E-012	-1E-025 -8E-025	5.79E-038
Cubic	0.411	204.333	3	811	0.000	21.714	1.06E-012		
S	0.094	84.024	1	813	0.000	3.113	-9E+007		
Growth	0.213	219.611	1	813	0.000	1.099	2.75E-014		
Exponential	0.213	219.611	1	813	0.000	22.184	2.75E-014		

Note: Independent variable is net sale

Estimation Model's Parameters

Total of Assets

As it said in abstract and according to the data nature can use panel analysis to fit the model. The reviewed model is:

$$Lny_{it} = B_0 + B_1 Lnx_{it} + e_{it}$$
$$I = 1, ..., n$$
$$T = 1, ..., T$$

Meaningful model in zero hypothesis and alternative hypothesis is:

H0= model is not meaningful.

H1= model is meaningful.

The amount of related responsibility to F statistic is equal to 0.000. This amount which is lower than 0.05 shows that the null hypothesis is reject in confidence level of %0.95. It means that the value of R2 of model is equal to

0.41 which in activity is considerable value. The only worrying sign in this model is the amount of Wattson Camera Statistic that the value of it is 0.35. This value shows remaining autocorrelation. So must do to solve autocorrelation. One of the ways to solve autocorrelation is using auto regressive or first grade AR.

Table 7. Variables Entered/Removed b

Variables Entered/Removed   b

Model	Variables Entered	Variables Removed	Method
1	Total assets log a	0.00	Enter

a. All requested variables entered

b. Dependent Variable: Total political costs log

Table 8. Dependent Variable

Prob.	t-Statistic	td. Error	Coefficient	Variable

Dependent Variable: LY
Method: Pooled Least Squares
Sample 2005-2007
Included observations: 3
Total panel (unbalanced) observations 814
White Heteroskedasticity-Consistent Standard Errors & Covariance
Cross sections without valid observations dropped

Prob.	t-Statistic	td. Error	Coefficient	Variable
0.0000	8.4697	8293	7.0242	C
0.0000	18.4819	0.0319	0.5903	LX1
22.4716	Mean dependent variance		0.8148	R-squared
1.2665	S.D. dependent variance		0.4141	Adjusted R-squared
763.1338	Sum squared		0.9694	S.E. of regression
0.3498	Durbin-Watson		575.5667	F-statistic
			0.0000	Prob (F-statistic)

By adding AR (1) part, have been improved the amount of Wattson Camera Statistic (the amount of this statistic can be expressive of naught autocorrelation of remaining if it be near to number 2.

This model is desirable model. The value of R2 in this model is 0.83. This model can be written like this:

$$Lny_{it} = 20.77 + 0.49\ Lnx_{it} + (0.9277AR\ (1))$$

Table 9. Method: Pooled

	Method: Pooled Least Squares			
	Sample: 2005-2007			
	Included observations: 3			
	Total panel (unbalanced) observations 469			
	Convergence achieved after 7 iterations			
	White Heteroskedasticity-Consistent Standard Errors & Covariance			
	Cross sections without valid observations dropped			
Prob.	t-Statistic	Std. Error	Coefficient	Variable
0.0000	7.6482	2.9269	20.7708	C
0.0000	9.0893	0.0539	0.4903	LX1
0.0000	36.5456	0.0234	0.9277	AR(1)
22.6239	Mean dependent variance		0.8355	R-squared
1.2543	S.D. dependent variance		0.8348	Adjusted R-squared
121.1202	Sum squared		0.5098	S.E. of regression
1.6403	Durbin-Watson		1183.4020	F-statistic
			0.0000	Prob(F-statistic)

The Amount of Sale

Meaningful model is confirmed such as the previous model because the probability of F is equal to 0.000. It means that there is a meaningful model between total sale and total political costs. The value of R^2 is equal to 0.64 but in this model the amount of Wattson Camera Statistic is low and its amount is 0.25. Perforce should use the AR (1) part in this model.

Table 10. Variables Entered/Removed b

Variables Entered/Removed b		
Entered	Removed	Method

a. All requested variables entered
b. Dependent Variable: Total political costs log

	Dependent Variable: LY			
	Method: Pooled Least Squares			
	Sample: 2005-2007			
	Included observations: 3			
	Total panel (unbalanced) observations 815			
	White Heteroskedasticity-Consistent Standard Errors & Covariance			
	Cross sections without valid observations dropped			
Prob.	t-Statistic	Std. Error	Coefficient	Variable
0.0000	4.8195	0.6658	3.2088	C
0.0000	28.6234	0.0261	0.7470	LX2
22.4720	Mean dependent variance		0.6415	R-squared
1.2658	S.D. dependent variance		0.6410	Adjusted R-squared
467.6032	Sum squared		0.7584	S.E. of regression
0.2481	Durbin-Watson		1454.5410	F-statistic
			0.0000	Prob(F-statistic)

Table 11: graph

Prob.	t-Statistic	Std. Error	Coefficient	Variable
	Dependent Variable: LY			
	Method: Pooled Least Squares			
	Sample: 2005-2007			
	Included observations: 3			
	Total panel (unbalanced) observations: 470			
	Convergence achieved after 3 iteration(s)			
	White Heteroskedasticity-Consistent Standard Errors & Covariance			
	Cross sections without valid observations dropped			
0.1221	1.5489	2.7561	4.2690	C
0.0000	7.2309	0.1006	0.7274	LX2
0.0000	25.6493	0.0342	0.8759	AR(1)
22.6232	Mean dependent variance		0.9173	R-squared
1.2533	S.D. dependent variance		0.9176	Adjusted R-squared
60.8977	Sum squared		0.3611	S.E. of regression
2.3904	**Durbin-Watson**		2591.0050	**F-statistic**
			0.0000	Prob(F-statistic)

The amount of probability in the model within AR is 0.000 which shows the meaningful model.

Table 12: Dependent Variable: LY

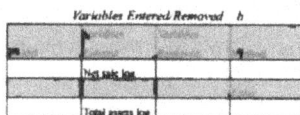

	Variables Entered/Removed b		
	Net sale log		
	Total assets log		

a. All requested variables entered
b. Dependent Variable: Total political costs log

Prob.	t-Statistic	Std. Error	Coefficient	Variable
	Dependent Variable: LY			
	Method: Pooled Least Squares			
	Sample: 2005-2007			
	Included observations: 3			
	Total panel (unbalanced) observations: 814			
	Convergence achieved after 3 iteration(s)			
	White Heteroskedasticity-Consistent Standard Errors & Covariance			
	Cross sections without valid observations dropped			
0.0000	4.8912	0.3153	1.4999	C
0.0000	-1.5995	0.0054	-0.0073	LX1
0.0000	21.3689	0.0376	0.8040	LX2
22.4239	Mean dependent variance		0.6452	R-squared
1.2665	**S.D. dependent variance**		0.6423	**Adjusted R-squared**
365.3816	Sum squared		0.7575	S.E. of regression
0.2301	**Durbin-Watson**		730.8632	**F-statistic**
			0.0000	Prob(F-statistic)

The value of R^2 increases to 0.92. It means that there is a very powerful relation between discussed variable and independent variable. Estimated model is:

$Lny_{it} = 4.26 + 0.72 \, Lnx2_{it} + [0.87 \, AR \, (1)]$

Multi Models

After fitting the model lonely in previous part it is needed that to estimate bottom multi model:

$$Lnyit = B0 + B1Lnx1it + B2Lnx2it + eit$$

The advantage of said model in comparison with simple line models is that it can calculate the net effect of each one by controlling other variables.

The null hypothesis and alternative hypothesis in this model for meaningful model is:

H0= there is not meaningful model H0:   B1 = B2 = 0

H1= there is meaningful model H1:   they are not zero at the same time.

The amount of F probability is equal to 0.000. This is as rejecting zero hypothesis or meaningful model. The amount of Wattsan Camera Statistic is equal to 0.25 which shows the autocorrelation of remaining.

Table 13: Dependent Variable: LY

Dependent Variable LY				
Method: Pooled Least Squares				
Sample: 2005-2007				
Included observations: 3				
Total panel (unbalanced) observations: 469				
Convergence achieved after 4 iteration(s)				
White Heteroskedasticity-Consistent Standard Errors & Covariance				
Cross sections without valid observations dropped				
Prob	t-Statistic	Std Error	Coefficient	Variable
0.0540	1.9316	2.6926	5.2012	C
0.3732	-0.8877	0.0555	-0.0493	LX1
0.0000	6.9690	0.1067	0.7433	LX2
0.0000	26.2334	0.0334	0.8762	AR(1)
22.6239	Mean dependent variance		0.9179	R-squared
1.798	S.D. dependent variance		0.9174	Adjusted R-squared
60.4329	Sum squared		0.3605	S.E. of regression
2.3846	Durbin-Watson		1733.4500	F-statistic
			0.0000	Prob(F-statistic)

Table 14: Variables Entered/Removed and Model Summary

Variables Entered/Removed

Year	Model	Variables Entered	Variable Removed	Method
2004	1	Total assets log, Net Sale log	.	Enter
2005	1	Total assets log, Net Sale log	.	Enter
2006	1	Total assets log, Net Sale log	.	Enter

a. All requested variables entered

b. Dependent Variable: Total political costs log

Model Summary

Year	Model	R	R Square	Adjusted R Square	Std. Error of the Estimate
2004	1	.796 a	.633	.631	.75846
2005	1	.792 a	.625	.625	.73917
2006	1	.815 a	.661	.661	.76098

a. Predictors: (Constant), Total assets log, Net Sale log

So by increasing AR part can increase the amount of Wattson Camera Statistic and get a valid model.

Table 15: Anova

Anova

Year	Model		Sum of Squares	df	Mean Square	F	Sig.
2004	1	Regression	325.695	2	162.847	283.082	.000 a
		Residual	188.687	328	.575		
		Total	514.382	330			
2005	1	Regression	248.027	2	124.013	226.851	.000 a
		Residual	147.055	269	.547		
		Total	395.082	271			
2005	1	Regression	238.278	2	119.139	205.735	.000 a
		Residual	120.451	208	.579		
		Total	358.729	210			

a. Predictors: (Constant), total assets log, Net sale log

b. Dependent Variable: Total political costs log

The value of R^2 is equal to AR part and increase to 0.92 and the model is meaningful, but X_1 variable is not meaningful. It seems that the X_1 and X_2 behavior is the same at relation with Y, but the value of X_2 relation with Y is more intensive than X_1 with Y. so in presence X_2 (sale) variable X_1 (total assets) is not meaningful. The estimated model can be written like this:

$$Lny_{it} = 5.2 - 0.049\, Lnx_1 + 0.74\, Lnx_2 + (0.87\, AR)$$

Fitting the Multi Model for Each Year

Previous said model can estimate by using regression analysis for each year. Considering to that data in this model is as periodic so doesn't have autocorrelation problem. In bottom table has brought meaningful model and amount of R2 and estimate parameters.

A. Regression

Table 16: Coefficients chart

D1. Median

		Frequency	Percent	Valid Percent	Cumu-lative Percent
Valid	.00	411	38.8	50.0	50.0
	1.00	411	38.8	50.0	100.0
	Total	822	77.6	100.0	
Missing	System	237	22.4		
Total		1059	100.0		

Table 17: Statistics

Statistics

	N			
	Valid	Missing	Mean	Median
Total assets	d 822	237	633313520217	214260762829
Net sale	826	233	412687441000	154824970679

For each year can write the model like this:

Table 18: D1. Median

D1. Median

		Frequency	Percent	Valid Percent	Cumu-lative Percent
Valid	.00	411	38.8	50.0	50.0
	1.00	411	38.8	50.0	100.0
	Total	822	77.6	100.0	
Missing	System	237	22.4		
Total		1059	100.0		

Estimate Models for Small and Large Companies

To dividing companies into small and large and survey the amount of effective sale and total assets in political costs have been used superstition (figurative) variables.

Table 19: D2. Median

D2. Median

		Frequency	Percent	Valid Percent	Cumulative Percent
Valid	.00	413	39.0	50.0	50.0
	1.00	413	39.0	50.0	100.0
	Total	826	78.0	100.0	
Missing	System	233	22.0		
Total		1059	100.0		

Table 20: graph

Model	Variables Entered	Variables Removed	Method
	D1med.x1, Total assets log , D1. a median		. Enter

a. All requested variables entered.

b. Dependent Variable: total political costs log

Table 21: graph

	Dependent Variable: LY				
	Method: Pooled Least Squares				
	Sample: 2005-2007				
	Included observations: 3				
	Total panel (unbalanced) observations: 467				
	Convergence achieved after 5 iteration(s)				
	White Hetero...istent Standard Error & Covariance				
	Cross sections without valid observations dropped				
Prob.	t-Statistic	Std. Error	Coefficient	Variable	
0.0000	10.4349	2.3804	24.8397	C	
0.7024	-0.3823	0.0819	-0.0313	LX1	
0.0756	1.7809	3.1048	-5.5291	D1	
0.0766	1.7746	0.1196	0.2123	LXILD1	
0.0000	37.6489	0.0245	0.9241	AR(1)	
22.6089	Mean dependent variance		0.8416	R-squared	
1.2543	S.D. dependent variance		0.8403	Adjusted R-squared	
112.6738	Sum squared		0.4917	S.E. of regression	
1.6611	Durbin Watson		613.8506	F-statistic	
			0.0000	Prob(F-statistic)	

This variable use if considered the type of relation in two or multi level, so D1 and D2 define like this:

D1= 0 if the company is small (according to total assets)

1 if the company is large (according to total assets)

D2=0 if the company is small (according to total sale)

1 if the company is large (according to total sale)

Table 22: graph

Dependent Variable: LY				
Method: Pooled Least Squares				
Sample: 2003-2007				
Included observations: 3				
Total panel (unbalanced) observations 470				
Convergence achieved after 5 iteration(s)				
White Heteroskedasticity-Consistent Standard Errors & Covariance				
Cross sections without valid observations dropped				
Prob	T-Statistic	Std. Error	Coefficient	Variable
470	1.8240	4.1737	7.6130	C
0.0003	3.6849	0.1598	0.5890	LX2
0.0451	-2.0087	4.4949	-9.0289	D2
0.0451	29.4187	0.1744	0.3503	X2D2
0.0000	37.6489	0.0302	0.8894	AR(1)
22.6252	Mean dependent variance		0.9213	R-squared
1.2533	S.D. dependent variance		0.9206	Adjusted R-squared
57.9708	Sum squared		0.3531	S.D. of regression
2.3223	Durbin-Watson		1360.9510	F-statistic
			0.0000	Prob(F-statistic)

$Lny_{it} = 7.61 + 0.58\,Lnx_1 - 9.02D_2 + 0.35Lnx_2D_2 + [0.89AR\,(1)]$

So multi lonely models are estimated like this:

$Lny_{it} = 24.8 - 0.7\,Lnx_1 - 5.52D_1 + 0.21Lnx_1D_1 + [0.92AR\,(1)]$

$Lny_{it} = 7.61 + 0.58\,Lnx_1 - 9.02D_2 + 0.35Lnx_2D_2 + [0.89AR\,(1)]$

Conclusion

The research based on this hypothesis that there is a correlation between political costs and company's size. It means that by increasing company's size the political costs increase too and by decreasing companies' size their political costs decrease too. Considering done analysis and get result which say spacious can say that there is a meaningful relation between company's size and political costs. While cannot say that the only factor that is determinant company's size is political costs (with scales sale evaluation or company's assets) and is undeniable the effect of other factors such as total politics dominant on market complex and the type of industry on political costs.

With regard to the results of this study the authors came to conclusions in the four areas which namely as follows:

A) Total of Assets

In this section considering to done analysis, get results and finally reject null hypothesis can say that there is a meaningful relation between company's size (assets volume) and political costs. In fact this result is explanatory grace number one hypothesis. Meantime by comparative comparison purposes can know this consonant deduction with get result in done research by Ayazi (2006) which is "there is a correlation between political costs and assets in large and small companies. According to it accept that assets volume affect all companies' political costs. It means that political costs are a function of companies' assets volume and there is correlation between company's size and political costs."

B) The Amount of Sale

Get results according to done analysis about the model which is based on the amount of sale (as an independent variable) is explanatory existence meaningful relation between company's size and political costs. In this section after rejecting zero hypotheses, number two hypothesis is graced based on existence meaningful relation between political costs and the amount of companies' net sale. This deduction is similar to the result of Firouzi's research (1998). In this research has shown that there is a close correlation between company's size at sale and political costs in each one of sections 1995 to 1997 and total relation too.

C) Multi Models

After fitting the model as alone and introduce above results it is needed to introduce multi models in order to calculate the net effect of each independent variable (assets volume and sale). Considering rejecting null hypothesis, results are explanatory meaningful multi model. It seems to be same the X_1 and X_2 behavior in relation with Y. but X_1 is not meaningful and notwithstanding two independent variables have same behavior of the amount of X_2 relation with Y (sale and political costs) is stronger than X_1 with Y (assets volume and political costs). So in presence X_2, X_1 is not meaningful. Done analysis in seventh part of section 4 which have done severance each year get us to above result. Introduced result in this section is similar to said result by Rahim Firouzi (1998). In his research during grace existence meaningful relation between each one of sale and assets variables with political costs are mentioned to this subject that severity of relation and correlation between assets and political costs in above period was not in the same size and viewed severity about sale and political costs.

D) The Results by Separating Companies into Small and Large

To introduce the model which includes small and large companies at the same time has used superstition (figurative) variables. Introduced analysis of model and related tables that its dependent variable are companies' assets volume show that there is not meaningful difference between models and small and large companies' coefficients. It means that whatever political costs are function of increasing assets volume (growing of companies) but regression model and coefficient of correlation don't have meaningful difference between dependent and independent variables in small and large companies. While get results of model analysis which is based on the amount of sales as a company size index is explanatory of existence meaningful difference between models and small and large companies' coefficient. It means that notwithstanding meaningful relation between dependent and independent variable in small and large companies, severity of relation and the amount of correlation is different with them. In fact severity of this relation in large companies is more than smaller. Whatever in Ayazi study (2006) is not used superstition variable to dividing companies into small and large but introduced results in this research are explanatory that despite establishing meaningful relation between small and large companies has been considered differences between models and small and large companies' coefficients.

Suggestion for Decreasing Political Costs

Now, after the appearing close correlation between company's size (especially at sale) and political costs introduce this question that what solutions should we take to control and decrease this type of costs and go toward more efflorescence country's trade units. The considerable point is the large part of country political costs such as tax, insurance, urban toll, education toll and ... suffer to companies with legal bankroll and its suffering is inescapable from companies, but another part of said costs like the cost of establishing celebrations, election costs, managerial munificence and help to election is controllable and can validate and decrease them by correct management doing exact auditing.

Suggestion for Future Researches

During writing this research and survey the relation between company's size and political costs and studying several essays about research subject or related subjects to political costs seem several subjects to research that some of them is:

1. Survey the relation between political costs and company's size according to the type of industry as comparison.

2. To determine company's size has been used another factors such as number of staffs and the number of stockholders and survey its relation between it and political costs.

3. Survey the effects of political costs on companies' managers' motivation in selecting accounting method.

4. Survey the effects of political costs on companies' divulgences process (management).

5. Survey the relation between company's size and effective rate of tax on companies.

6. Survey political costs and dept treaties.

7. Survey how encounter independent auditor with political costs subject and their solutions to decrease or delete this type of costs.

Research's Limitations

The process of doing each research considering to its nature in encounter to a series of limitations which cause slow working or naught necessary clear and total deduction of it. Some of the available total limitations are:

1. Lack of translated resources in country (getting to external resources was not possible easily).

2. Lack of related research to research's subject in the country or out of it.

3. Hiding the numbers and digits of political costs by some of the companies because of subject's sensitivity.

4. The related problems to getting to financial statements information, especially political costs things.

References

Ball,R.J., & Foster, G (1982), Corporate Financial Reporting, Journal of Accounting Research, Vol. 20, pp. 161–234.

Bazargan, Abbas; Sarmad, Zohreh and Hejazi, Elaheh (2000), Research Method in Behavioral, Agah publisher, p. 20–230.

Bekaoui,A. & Karpik, P.G (1989), Determinants of the Corporate Decision to Disclose Social Information, Accounting, Auditing & Accountability Journal, Vol.2, No.1, pp. 85–107.

Blacconiere,W.G. & Patten, D.M. (1994), Environmental Disclosures, Regulatory Costs, and Changes in Firm Value, Journal of Accounting and Economics, Vol. 18, pp. 357–377.

Bozorg Asl, Mousa (1994), The Role of Debt Treaties, Gratuity Plan and Political Process in Selecting Accounting Rules, Iranian Accounting and Auditing Journal, pp. 6–7.

Christopher, C. Ikin (2003), Political Cost Influences on the Determination of Non-Audit Services, The Australian National University-Canberra-Www.Ecocomm. and.edu.au.

Deegan,C & A Hallam, (1991), The Voluntary Presentation of Value Added Statements in Australia: a Political Cost Perspective." Accounting and Finance, 31(1), pp. 1–21.

Firouzi, Rahim (1998), Survey Political Costs in Mazandaran Stock Exchange: Solution to Reduce this Cost, MS Dissertation, Mazandaran University.

Godfrey, J. & K Jones, (1999), Political Cost Influences on Income Smoothing via Extraordinary Item Classification, Accounting and Finance, 39(3), pp. 229–253.

Healy, P (1985), The Impact of Bonus Schemes on the Selection of Accounting Principles, Journal of Accounting and Economics, Vol. 7, April, pp. 85–107.

Holthausen,R.W. & R.W Leftwich, (1983), The Economic Consequences of Accounting Choice: Implications of Costly Contracting and Monitoring," Journal of Accounting and Economics, Vol. 5. pp. 77–117.

Hosseinzadeh, Sha'ban (2003), The Relation between Company's Size and Political Costs in Rah Gostar Developing International Trade company, MS dissertation, Mazandaran university.

Houman, Heidar (1994), Base of Research Method in Behavioral, Parsa publication, Third edition, p. 167.

Jerry,C.Y. Han & Shiing, Wang (1998), Political Costs and Earnings Management of Oil Companies during the 1990 Persian Gulf Crisis, Accounting Review, Vol. 73, 103–117.

Kahan, S (1992), The Effect of Antitrust Investigations on Discretionary Accruals; a Refined Test of the Political Cost Hypothesis, Accounting Review, Vol. 67, pp. 77–95.

Khaki, Gholamreza (2004), Research Method in Management, Azad Islamic University Publication Center, Third edition, pp. 188–189.

Markus,J.Milne, (2001), Positive Accounting Theory, Political Costs and Social Disclosure Analyses: a Critical Look.

Panchapakesan,S. & J .McKinnon, (1992), Proxies for Political Visibility: a Preliminary Examination of The Relation among Some Potential Proxies, Accounting Research Journal, Spring, pp. 71–80.

Patten, D. M. (1991), Exposure, legitimacy and Social Disclosure, Journal of Accounting and Public Policy, Vol. 10, pp. 297–308.

Pourheidari, Omid. Hemmati, Davoud (2004), Survey Debt Treaties, Political Costs, Gratuity Plan and Ownership on Management Profit in Tehran Stock Exchange Listed companies, Iranian Accounting and Auditing Journal, Vol. 36, pp. 47–63.

Seifollahi, Seifollah (2004), Political Economic, Essays and Plans Complex, Almizan Publication.

Shabahang, Reza (2004), Accounting Theory, Audit Organization Publication., Third edition.

Takzare, Nasrin (2001), Guide to Dissertation Writing, Report of Research and Essays, Teymourzadeh publication, 2nd edition.

Watts, R. (1997), Corporate Financial Statements, a Product of Market and Political Processes, Australian Journal of Management, Vol. 2, pp. 53–75.

Watts,Ross.L. & Jerold. L. Zimmerman (1986), Positive Accounting Theory.

Wong. J. (1998), Political Costs and an Intra period Accounting Choice for Export Tax Credits, Journal of Accounting & Economics, Vol. 10, pp. 11–19.

Yangsen Kim; Caixing Liu & S. Ghon Rhe (2003), The Relation of Earning Management to Firm Size, Journal of Management Research, Vol. 4, pp.81–88.

Zimmerman. J.I (1983), Taxes and Firm Size, Journal of Accounting and Economics, Vol. 5, pp. 119–149.

An Evaluation of Inter-Organisational Information Systems Development on Business Partnership Relations

Elizabeth A. Williamson

ABSTRACT

Inter-organisational information systems (IOS) are being used within SCM to improve businesses processes and to facilitate closer working relations with business partners. However, the technologies themselves impact on this relationship as they allow various levels of information flows, communications, function integration and partner integration.

The aim of this paper is to evaluate IOS development influencing partnership integration within Supply Chain Management (SCM) by investigating thirteen businesses that use a variety of IOS. IOS are classified into Elementary IOS, Intermediate IOS and Advanced IOS. Organisational variables such as information flows, partner co-ordination and integration, partner trust and confidence are measured against the level of IOS development. Variables such as management commitment, financial costs, system standards and partner resistance are investigated as forces or barriers, and related to different levels of IOS development.

This research concludes that IOS development results in increased information flows and coordination which supports the development of trust and confidence in business partners. However, the customer position in the supply chain, whether it be retailer, distributor or manufacturer can influence the use of IOS of its business partners. Also, although IOS allows businesses to source and contact a larger range of business partners, the tendency is for businesses to use a smaller number of partners. This work also shows that IOS development changes a business's relationship with its partners and moves it towards partnership integration.

However, a number of organisational factors impact on this integration. These factors vary with the level of IOS development. Management commitment and showing the requirement for IOS development can act as a positive force in developing IOS or as a barrier against IOS development. Other barriers include resistance from business partners, financial costs, lack of system standards and technical maturity of the companies. The effect of these barriers also is affected by the level of IOS development.

Keywords: inter-organisational information systems, supply chain management, business partnership relations

Introduction

An inter-organisational information system (IOS) is a collection of IT resources, including communications networks, hardware, IT applications, standards for data transmission, and human skills and experiences. It provides a framework for electronic cooperation between businesses by allowing the processing, sharing and communication of information (Haiwook, 2001). IOS are also known as extranets and allow electronic processing of business transactions and documents, as well as the transfer of information with minimal effort and makes it quickly available. The growth of IOS in SCM has allowed the flow of information throughout the supply chain by the integration of business processes (Stephens, Gustin & Ayers, in Ayers, 2002).

IOS can be used at both ends of the supply chain. They can be used with customers, to give visibility of data and interaction with company employees and with business partners, such as suppliers and logistic companies. Benefits are the same at both ends and include visibility of data and reduced purchasing costs (Ayers, 2002).

IOS can be categorised into four phases, in terms of historical IS development (Shore, 2001):

- Phase One—Manual systems
- Phase Two—EDI systems
- Phase Three—ERP systems
- Phase Four—Internet-enabled systems

Phase One: Manual Systems

This phase includes paper copies of documents such as purchase orders, bills and invoices. The information is processed manually and therefore information technology and telecommunications do not contribute to this system. The disadvantages of this phase are obvious—laborious procedures, inaccurate data, insufficient information and expensive maintenance of the system. This initial phase is still in use in some companies either on a wide scale across the company or in particular departments and in many Small to Medium sized Enterprises (SMEs), as shown in the empirical case studies. This IOS development may be curtailed due to a lack of expertise, financial resources or other organisational or environmental pressures (Papazoglou & Ribbers, 2006).

Phase Two: Electronic Data Interchange Systems

The next phase involved the development of EDI technology in the 1980's and this had a dramatic effect on the automation of heavy data flows and the elimination of many labour intensive key business processes. Paper documents such as purchase orders, invoices, bills of lading and shipping slips were replaced by electronic transmission of the information between computers (McKeown, 2003). EDI was the main technology used in electronic trading in many sectors, such as retail, manufacturing and financial services and has only been widely used since the early 1990's (Willia ms & Frolick in Ayers, 2002).

Early EDI systems used value added networks (VAN), which are special services on public networks available by subscription, and provide companies with data communication facilities. The company operating the VAN is totally responsible for managing the network, including providing any data conversion between different systems (McKeown, 2003). Therefore VANs were expensive to implement and therefore limited EDI use to the larger companies.

However, as there is no single agreed national or international standard, an EDI link tends to be set up for a specific supplier and buyer. Therefore it is difficult to switch the connection to another partner and a second EDI system may have to be created. Golden and Powell (1999) research showed that EDI limits flexibility of suppliers who are connected to more than one customer since they are required to support specific technologies for each. This resulted in an explosion of EDI software, VAN and EDI standards which made it difficult to integrate the technologies. Also, the full benefits of EDI can only be realized when EDI is fully integrated with other transaction processing systems, such as accounting and sales systems. The information sent via EDI is ordered as a transaction set and this transaction set has a fixed structure. These transaction sets for new products or services also have to be firstly agreed before they can be implemented. These data constraints also hamper EDI growth and implementation.

A second generation of EDI technology, Internet EDI, overcomes some of the disadvantages of the early EDI systems. Companies are able to use existing EDI systems and processes by installing Extensible Mark-up Language (XML) EDI translators on web-servers. Internet EDI lowers entry costs for businesses as data is transmitted over the Internet rather than using subscription to a VAN, and therefore telecommunication costs are minimized. It is also more useful in the global marketplace. Cost savings can be as much as 90% (EDI Data, 2003) and therefore can be implemented by smaller companies. It uses the same EDI standards for documents. Data transaction sets are also more flexible within Internet EDI and allows easier and quicker development of applications (Papazoglou & Ribbers, 2006). Data is processed in real-time when using Internet EDI, as opposed to overnight batch data flows/processing and this is also an operational advantage. Therefore, due to these benefits over the older system, the volume of Internet EDI is increasing.

Phase Three: Enterprise Resource Planning Systems

This phase describes a more integrated information systems approach. This approach is being taken by companies who view the integration of systems and information flows as being essential in providing improved customer satisfaction and cut operational costs in an increasingly competitive market-place (Jenson & Johnson in Ayers, 2002).

Enterprise-wide systems and databases integrate and coordinate IT operations across the company. These systems, characterized by Enterprise Resource

Planning (ERP) systems, have developed from Manufacturing Resource Planning (MRP II) applications. They generally include manufacturing, logistics, distribution, inventory, shipping, invoicing and accounting (Ayers, 2002). The integration of information from all departments in the company in the ERP system means that output or consequences from one system can be fed into other systems, so that there is total information coordination. An ERP system can also assist in controlling business activities such as sales, delivery, billing, production, inventory management and human resource management. Therefore it can cover all primary and support activities within the Value Chain. The implementation of ERP systems also results in organisational efficiencies as they automate processes, integrate functions and improve the quality of information flows (Papazoglou & Ribbers, 2006). The reach of ERP systems can be extended to include partners with the supply chain by the use of SCM software transferred onto the new integrated system. ERP systems, such as SAP's R/3, have been implemented across the globe. Worldwide sales of ERP packages, combined with implementation support, exceeded $15billion in 1999 with annual growth rates of over 30% (Akkerman et al., 2003).

Phase Four: Internet-Enabled Systems

The Internet is a worldwide web of computer networks. The development of the protocol, Transmission Control Protocol/Internet Protocol (TCP/IP), allows separate networks of different architectures to work together through open network architecture. The integration of information resources has therefore been enabled by the use of web development technologies such as Extensible Mark-up Language (XML) and Java, which have allowed business partners to integrate their information resources. These systems also provide platforms for fast and reliable communications between trading partners, regardless of physical barriers (Bandyo-padhyay, 2002).

The use of the Internet requires integration of computer systems by examining existing legacy systems and software and developing integrated solutions. However, changing corporate information systems brings about a number of challenges for the business which need to be managed successfully (Krizner, 2001):

1. Businesses have invested thousands if not millions of pounds in legacy systems which they will be keen to keep in place

2. The financial and time resources required to carry out systems integration

3. Security and risk aspects of opening up internal systems to external parties

4. Legacy systems require to be integrated to allow information flow between disparate systems and were not designed to 'talk' to other systems

5. Businesses may define processes and data differently from their supply chain partners

6. Legacy systems of partners may use different platforms

7. Partners will belong to many different supply chains

There are a variety of information mechanisms available for use by managers in SCM, such as auctions, purchasing groups and electronic agents which provide this linkage. Recent developments also include trading exchanges or market places. These are online supply chains which allow the sharing of real-time synchronized information by using XML on features such as prices and delivery information. Examples include NonstopRX.com for the pharmaceutical industry and Retail.com for apparel manufacturers and buyers (Messmer, 2000). These mechanisms may be used to conduct a business transaction, to purchase something at a given price or to share information to coordinate the flow of the item after the purchase has taken place. These collaborative mechanisms come under the Collaborative Planning, Forecasting and Replenishment (CPFR) heading and aim to closely integrate business partners. In order to help companies come together within this system, the Voluntary Inter-industry Commerce Standards Association (VICS) publishes guidelines to assist companies to achieve their objectives when using CPFR systems. However, the technology may require some business process change and also CPFR should be integrated into the e-business strategy. (Grossman, 2004). Managers are also required to choose the appropriate level of integration for particular relations in the supply chain and the appropriate degree of information sharing (Garcia-Dastugue & Lambert, 2002).

Therefore the Internet is now being used as one of the main networking platform in the upstream, downstream and internal supply chain by both large and relatively small companies.

Current IOS Development within Business

However, companies may be involved in ad-hoc development and use various operational and management information systems. For example, a company may use a legacy system for processing orders and stock control. Legacy systems may also be 'best of breed,' bespoke, point or developed internally ERP systems. It may access a customer's web-enabled production system to calculate order quantities and it may implement a new invoicing and accounting system as required by head office which may be a SAP system. These systems may also be totally disjointed, totally integrated, using a web-enabled ERP integrated technology or use a few information systems with some integration of processes, some of which are web-enabled. Such companies may have the objective of full integration in the future. For example, a substantial proportion (82%) of companies surveyed during 2003 expect to be using the web for purchasing in the course of the next few years, while 75% plan to use online technology for order management and 71% for order status, with supplier management (69%) and selling (65%) proving equally popular. (Sweet, 2003). Therefore the actual configuration of IOS used by a company usually consists of a number of IOS, which may be partially or fully integrated internally or externally with business partners. This scenario is evident in the empirical case studies.

Inter-Organisational Information Systems Development Affecting Partnership Relations

Previous research has focused on the how the use of information systems themselves have changed the business structure and influenced partner relationships (Venkatramen, 1991, 1994). Christopher & Juttner (1998) found that the quality of a relationship is strongly influenced by its interface structure. Premkumar in Ayers (2002) states that the nature of the IOS technology and partner linkages, including common partner objectives is important to IOS development and implementation.

Electronic Data Interchange Systems and Partnership Relations

Electronic Data Interchange (EDI) and point of sale (POS) systems have been used generally within SCM to facilitate information flows and therefore communication between businesses and their partners. Research carried out by Hill and Scudder (2002) on the use of EDI systems in the food industry found that the implementation of EDI facilitated inter-company coordination and that EDI users have more coordination with their suppliers than do non-users of EDI. These IOS bring about electronic cooperation by integrating stock holding, distribution, purchasing and other functions to improve customer responsiveness (Mische, 1992). POS systems have been a major influence in increasing information sharing among, for example, logistic managers and show that information exchanges can be beneficial to all parties involved (Lancioni, Smith & Oliva, 2000).

Haiwook (2001) found that EDI provides a better means of inter-organisational coordination than earlier applications of IT, but it is limited in its influence. Similarly, Santema (2003) found that although EDI improve communication between business partners, it doesn't 'add value' to the relationship. This supported previous findings by Clemons & Row (1993), who found that EDI-based checkout scanners significantly affect efficiency and information flow in distribution channels, but that automating the processes in this way, doesn't increase electronic cooperation. He concluded that only by sharing company information, such as in a two-way information flow, is a true partnership developed. EDI has also been accused of depersonalizing inter-organisational relationships due to its restricted information format (Morris, Tasliyan & Wood, 2003).

Crook & Kumar (1998) found that a number of organisational variables, including the partners' experience of EDI influenced the expansion of its use. Vlosky et al. (1997) found that buyers, such as manufactures, initiate EDI and expect to and actual gain more benefits that their suppliers—their main objective is to improve customer satisfaction and cut costs. Significant disruption to the business relationship can occur if the implementation is not handled properly. Over time, relationship strength and satisfaction increases.

Internet Inter-Organisational Information Systems and Partner Relations

The Internet can be used to provide a platform for partnerships in all areas of the supply chain, whether it is procurement, purchasing, negotiation, coordination or just information exchange. The Internet al.lows two-way communications, unlike EDI technology, and therefore has much more impact on partner relations and partnerships. Research by Lancioni, Smith & Oliva (2000) concluded that the use of Internet IOS, can improve supplier relations by improving communications and data flows between suppliers and businesses. Support for these conclusions is given by a number of researchers. Barua et al. (2001) suggested that the Internet provides opportunities for companies to develop relations with all business partners, suppliers as well as customers. This research was further developed by Zank and Vokurka (2003) who surveyed manufacturers, distributors and industrial customers and found that overall, members of the supply chain believed that e-commerce had a slightly positive impact on their relations with other supply chain partners. Hayes (2002) has shown that the use of the latest IT systems can aid supplier relations. For example, commitment, trust and communications can be enhanced and solidified by allowing the supplier/s access to real-time data which can also be manipulated, as required.

The Internet better supports the business relationship by offering better information with little investment expense. (Papazoglou & Ribbers, 2006). Therefore is it the case that advances in technology are changing this relationship scenario and now it is not so much about the type of transaction but the technical development within companies and level of information interaction between companies.

Li & Williams (2001) concluded that implementing IOS could assist in strengthening partnerships and improving cooperation as

- it requires close working between companies which, in turn, helps them to build a closer relationship and encourage the sharing of information
- it removes many errors associated with manual systems
- changing partners can be costly and time consuming

However, some Japanese car manufacturers see e-procurement, in particular, as preventing the development of closer partnership-type relations, which is what they prefer, by automating partner relations and therefore limit e-purchases to nuts and bolts and basic office supplies (Harney, 2000).

Research

This paper further develops the above findings by investigating the influence of IOS development on business relationships with suppliers and customers. It also analysed forces for and against IOS development.

The research critically evaluates 13 case studies to provide rich, in depth information to develop a theoretical model, thereby extending existing theories and models. The case study approach was also chosen as an all-embracing method that allows a detailed investigation and understanding of situations within particular organisational settings (Walsham, 1995). Nineteen in-depth interviews were carried out with senior employees of these 13 companies to enable cross-case analysis. Companies in product supply chains were the focus of the research as the products, and supply chains would be clearly identifiable. Most of the companies were large multinational companies with locations in Scotland, although three were small Scottish Companies. Items produced by all companies included electronics, clothing, food and drink, packaging and fixing solutions. The companies were positioned at various points in the supply chains—suppliers, packaging distributors and manufacturers. Retailers were also included in the research in order to give an analysis of the full supply chains.

Initially, employees interviewed were managers responsible for part of the supply chain, such as purchasing managers and sales managers. Thereafter, the manager was usually asked to suggest partners, either suppliers or customers, as appropriate, who could be used for the development of dyadic case studies. Managers in these companies were interviewed by using the same structured questionnaire. Interview responses were confirmed and corroborated by managers within the same companies, partner companies and other companies using the same type of technologies. It was also substantiated by additional information from a number of secondary sources, such as company publications, company websites, published company case studies and the WWW. Data collected was made more robust by the supply chain network associations. For example, third-party systems, such as GXS TradeWeb, an Internet EDI global marketplace used by several companies included in the research.

As this research investigated the relationship between a company and its partner, partners of the companies chosen were also investigated where possible. Wilson, Stone & Woodcock (1996) suggested that researchers exploring buyer-supplier relationships in business-to-business markets should collect data from both ends of the dyad. They argued that the collected data would be richer and therefore would compensate for the smaller sample size involved. In fact, although some linkages between case studies were not known, during the research and analysis, it was clear that most companies had linkages with most of the others. These gave very strong basis for conclusions. The case studies can therefore be seen as a network of companies.

Data analysis was undertaken during and after the data collection. The analysis was conducted using principles of hermeneutics (Klien & Myers, 1999) as methods for identifying and extracting key themes from multiple case studies (Eisenhardt, 1989). Data analysis used coding, developing trends, summarizing, clustering and graphs.

After an initial analysis of data, IOS used by the companies were categorised into three types, Elementary IOS, Intermediate IOS and Advanced IOS, according to the level of internal information systems integration and external information systems integration with partners in their supply chain. This classification was subsequently used to determine the objectives of the research and is further explained in Figure 1.

Inter-organisational Information System	Level of Information System Integration
Elementary	Many different internal IS with manual data input between systems. No systems integration with external partners
Intermediate	More than one internal IS may be used but automatic data input between systems. No systems integration with external partners
Advanced	More than one internal IS may be used but automatic data input between systems. More than one external-facing IS may be used with partners but automatic data input between internal and external systems.

Figure 1: Inter-organisational Information System Categories and Level of Information System Integration

Findings

The empirical study gave interesting findings on how IOS development affects not only the company itself, but also its relationship with its business partners, whether suppliers and customers. This is due to the fact that IOS cannot be deemed as a closed system on its own, but must be taken as a component of the organisation (Leavitt, 1965: Boddy, Boonstra & Kennedy, 2002).

Company Restructuring, IOS Development and Position in the Supply Chain

Findings here showed that companies restructure when developing IOS, to improve relationships with both suppliers and customers. All companies using Advanced IOS had restructured to some ext ent, while around 66% of companies using Intermediate IOS and only 33% of companies using Elementary IOS had undergone any restructuring. Therefore, companies with Advanced IOS are more likely to have restructured than companies using Intermediate or Elementary IOS. This supports previous findings that relationships within SCM are now increasingly being seen as partnerships and businesses within the supply chain generally consider themselves as partners, rather than taking on a more adversarial and segregated role (Grieco, 1989: Kanter, 1994: Bowersox, 1996: Heikkila, 2002).

Similarly, Venkatramen (1994) in Papazoglou & Ribbers (2006) determined that integration of advanced information systems require business transformation, which involve changing business structure and processes and establishing inter-organisational business processes. His work is supported by other writers such as Ayers (2002), Clark & Stoddard (1996) and Benjamin et al. (1990) who propose that benefits from IOS can only be gained when basic organisational structures and work processes are redesigned.

This research also showed that companies are more likely to reorganise to improve relations with customers due to the position power of the customer in the supply chain. Research supports these findings (Ayers, 2002) in that where previously power resided with vendors/suppliers and they pushed technology onto their customers, the retailers, now the balance of power is with the retailer. This power balance was also shown here. For example, a retailer in this study was pushing the food manufacturer to used newer advanced technologies, such as RFID. Another manufacturer was 'invited' to use the third party GlobalNetXchange marketplace by their retailing business partner.

Development of Trust and Confidence in Partners

The research found that the level of IOS development has a beneficial impact on communication/information flows. Of companies using Elementary IOS, 63% of respondents found no change on communications/information flows/interaction, with only 37% experiencing a Positive Effect. With regard to companies using Intermediate IOS, 82% replies experienced at least a Positive Effect with 28% experiencing a Significant Positive effect. All companies using Advanced IOS found that they had a positive effect (55%), or a significant positive effect (45%) on communications.

To illustrate, one manager commented that their legacy IOS brought about improvements in communications in purchasing, logistics and customer service and warehousing when they were implemented. Similarly, another employee commented that IOS development has not changed the number of meetings with suppliers, but has allowed more interaction with partners and therefore more important matters can be discussed at face to face meetings. A retail manager illustrated the beneficial impact of IOS on communications/information flows to his role. Weekly information from Head Office is fed back to himself to give information on store targets such as revenue, stock levels and shelf space usage.

Information sharing and communication is necessary to build trust (Ballou, Gilbert & Mukherjee, 2000). Therefore given the previous findings that IOS development can improve information sharing and communication, it was of interest to determine if trust between partners was also developed when IOS development has taken place.

Findings showed that Elementary IOS can have a detrimental effect on trust levels between partners. For example, one manager regarded the lack of technology as hindering trust levels between partners as he felt that, as his company's IOS were not up web-based and in real-time, they hindered communication between business partners. Where companies used only Intermediate IOS then a positive effect on trust levels was reported by 54% respondents. For example, a Purchasing Manager reported that their web-enabled link with suppliers such as has enabled them to share more information and has led to an increase in trust between the companies. Most, 82%, of the replies from the companies with Advanced IOS reported a positive effect on the trust levels with partners. Therefore IOS development assists in development of trust levels between partners.

These findings can be shown as in Figure 2, as a virtuous partnership circle, supporting the partnership.

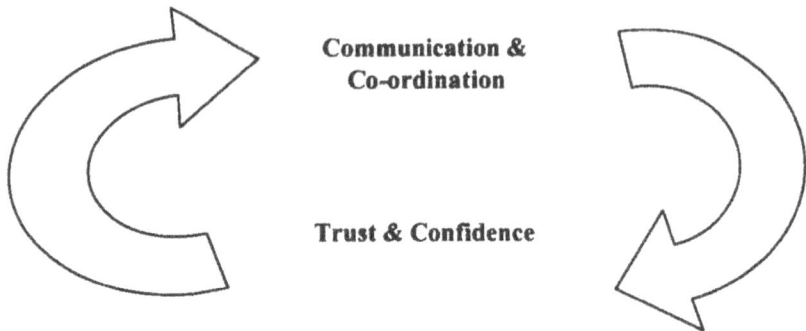

Figure 2: Virtuous Partnership Circle

Choi (1999) also found that there is a positive relationship between information volumes, amount of sales and joint decision-making, leading to better electronic cooperation. Grieco, (1989), Kanter (1994) and Kwon & Suh (2004) pointed to trust and communications/interaction as improving or even are necessary for effective working relationships with suppliers.

Barriers to and Forces for IOS Development

Findings showed that other organizational factors also impact on the rate of this development. Some of these factors can encourage IOS development, whilst others act as barriers against it.

Seven organisational factors were shown to be barriers—benefits not demonstrated, financial costs, lack of system standards, resistance from other business partners, resistance from customers, technical maturity of the company, and technical maturity of the trading partner. These barriers were found to have different levels of impact, depending on the level of IOS development. The barriers had least effect in companies using Elementary IOS and most impact in companies with Advanced IOS. This is supported by Soliman & Janz (2003) who found that EDI and Internet-based IOS encountered a range of barriers which varied with the level of IOS development.

'Benefits not demonstrated' and 'Financial Costs' were not significant barriers for Advanced IOS but tended to be significant barriers for Intermediate and Elementary Systems. The remaining five factors were shown to be barriers, but at different levels and for different levels of IOS development:

'Business requirement' and 'Management commitment' can act as positive forces and encourage IOS development at all levels, thereby also enhancing partnership relations. In fact, management commitment can act as both a barrier, if it does not exist and have positive impact if it does exist.

A manager within one of the electronic companies endorsed this finding with his comment that the implementation of sophisticated IOS allows both their business and their customer to benefit, thereby keeping them in a leading position in the electronic market.

Significant Barriers for Advanced IOS

- Lack of System Standards
- Resistance from other business partner
- Resistance from customer

Small Barriers for Advanced IOS

- Technical maturity of company
- Technical maturity of trading partner

Small Barriers for Elementary and Intermediate IOS

- Financial Costs
- Lack of System Standards
- Resistance from other business partner
- Resistance from customer
- Technical maturity of company
- Technical maturity of trading partner

Therefore, in relating the strength of barriers to the type of IOS, then problems were generally regarded as 'Small Barrier' with Elementary and Intermediate IOS and more as 'Significant Barriers' with regard to Advanced IOS.

The above analysis also shows that partner resistance is more important to companies with Advanced IOS. This supports other findings in that variables within both partners can influence partner integration and that partner variables are more important in companies deploying Advanced IOS.

Therefore, it may be that due to the sophisticated nature of the systems and the integrative nature of their deployment in bringing companies together, a stronger, 'leader' or champion for technological change is required in order to push through the Advanced IOS, and to overcome technical and partner barriers.

Model Development

In order to further progress the empirical and literature findings, a model has been developed from the above analysis. Firstly, this research has allowed the definitions of Elementary IOS, Intermediate IOS and Advanced IOS to be extended, to include other variables, such as Use of IOS, Partner Factors and Organisational Factors. Thereafter, the influence of organisational and technological factors at the three levels of IOS development on the organisation and its partners is compared. These are shown in the following Figures 3, 4, and 5.

Characteristics of Elementary IOS			
Technology	**Use**	**Partner Factors**	**Organisational Factors**
Many different IS used, including EDI and third party networks. No/little internal systems integration. Technology seen as an operational tool, rather than as a key strategic component. Automation of processes gives effectiveness gains.	Transactions only. Partners enter information into their own systems. Limited information communication and co-ordination. Limited benefits from use of technology. Collaboration at operational level.	Companies may be working towards their own agenda and for their own benefit. Trust exists between partners but is limited by the nature of the IOS used. Partner collaboration is weak. Customers exert position power in the chain	*Significant barriers:* Benefits not demonstrated *Small barriers:* Financial costs Lack of system standards Resistance from other business partners Resistance from customers

Figure 3: Characteristics of Elementary Inter-organisational Information Systems

Figure 3 shows that within Elementary IOS development, there is a low level of systems integration, information co-ordination, and partner collaboration and of benefits gained. Significant barriers are management commitment, benefits not demonstrated and resistance from customers. Smaller barriers include financial costs and resistance from other partners.

Figure 4 shows that there is some systems integration, information co-ordination, partner collaboration and benefits gained within Intermediate IOS development. Significant barriers and insignificant barriers seem to be similar to those in the Elementary IOS.

Characteristics of Intermediate IOS			
Technology	**Use**	**Partner Factors**	**Organisational Factors**
More that one IOS are used, but internal integration between information systems, gives integrated data flows. Role of technology is changing, from an operational tool to being a more strategic component. Technology is used for a larger range of tasks within all functions. Automation of processes gives effectiveness and efficiency gains. Information also used for business planning	Functional transactions and management reporting. Some integration of information systems to form links with parent company. Improved level of communications between functions. IOS replacing some face-to-face communications. IS supports communications with partners. Collaboration at operational and tactical levels	Companies may be working towards their own agenda and for their own benefit. However, some improvement in partner confidence due to increased communications and collaboration. Trust between partners is being enhanced. Integration of partnerships increasing Customers exert position power in the chain	*Significant barriers*: Benefits not demonstrated *Small barriers*: Financial costs Lack of system standards Resistance from other business partners Resistance from customers

Figure 4: Characteristics of Intermediate Inter-organisational Information Systems

Figure 5 shows the characteristics of Advanced IOS.

Characteristics of Advanced IOS			
Technology	**Use**	**Partner Factors**	**Organisational Factors**
Technology viewed as a key strategic component with information as a key resource. Internal and external integration between information systems, using one of more Internet systems and portals Automation of processes gives efficiency and effectiveness gains.	IOS are used at all levels within the org, from strategic through to operational. Technology is used for an extensive range of tasks within all functions. Integration of communications, functions and processes is carried out by sophisticated technologies	Advanced IOS allow partners controlled access to extensive company information Company may review the status and number of partners. They may use tiered partners according to 'value' of partnership. Customers exert position power in the chain	*Significant barriers*: Business Requirement Management Commitment Lack of System Standards Resistance from other business partner Resistance from customer *Small barriers*: Financial costs Technical maturity of company Technical maturity of trading partners

Figure 5: Characteristics of Advanced Inter-organisational Information Systems

At this level of IOS development, there is a high level of systems integration, information co-ordination, partner collaboration and benefits gained. Organisational barriers become different at this level of IOS development, with resistance from other business partners becoming more influential.

Combining these results into one graph, Figure 6 shows the influence of these organisational factors is the three levels of IOS development. The figure shows that IOS development impacts business partner relationship in a number of ways:

- Communication and coordination with the business partner increases

- Partner integration increases

- Confidence and trust in partners increase

However, this development also brings with it an increase in implementation barriers, such as lack of IOS standards and resistance from business partners. Management commitment and business requirement for IOS development can both as act as barriers against or drivers for IOS implementation.

Factor	Elementary IOS	Intermediate IOS	Advanced IOS	Factor Impact
Partner Communication co-ordination, integration				IOS development increases communication and co-ordination between partners, as well as developing partner integration
Partner confidence and trust				IOS development, enhances confidence and trust in partners and therefore partnership integration
System Standards				Impact of lack of information system standards increase in effect with IOS development
Resistance from Business Partners and Customers				Impact of resistance from other business partners and customers increase in effect with IOS development
Management Commitment				Top management beliefs can act as a barrier or driver for IS development at all levels of IOS development. Strong management commitment is required for significant IOS development
Business Requirement/ Benefits not demonstrated				Business requirement can act as a driver towards IS integration at all levels of IOS development. If the benefits of IOS implementation cannot be demonstrated, then this may cause a significant barrier

Figure 6: The Impact of Organisational Factors at varying levels of Inter-organisational Information Systems Development

Sherer (1995) developed a framework to describe three types of risk that affect IOS; technical risk such as security breaches, organisational risk which may arise due to restructuring of staff and their roles and environmental risk where partner and competitive forces exert influence over the company. Li & Williams (2001) also developed a similar three level model of barriers to the use of IOS, namely technical barriers or problems, organisational attitude to sharing information with business partners and suppliers and at the third level, an overall organisational cultural attitude towards inter-firm collaboration. However, this model relates such barriers to different levels of IOS development.

This model is supported by findings from Bensaou & Venkatramen (1995) who found that Inter-organisational Relationships varied with IOS development and that new IOS development will result in new business models and relationships. The pressure for these developments will come from the highly competitive marketplace.

Conclusion

This study is a comparative study on levels of IOS development within 13 case studies. It provides a better understanding of the impact of IOS across organizational boundaries. The findings show that the level of IOS development influences partner co-ordination and integration. IOS capabilities also assist in building trust and confidence in partners. However, a number of organisational factors influence IOS development and these factors such as management commitment, financial costs, resistance from other business partners can act as forces for or barriers to IOS development. The strength of these variables also varies with IOS capabilities.

The study has important implications for business. Organisations are increasing their use of IOS within SCM functions and therefore identification of the influencing factors is required and critical for emerging electronic business environments. Since partnership arrangements can be difficult and resource intensive and important to success in SCM, it would be valuable to businesses to evaluate the impact of technology, in particular, Internet IOS, on the required levels of partnership integration in particular SCM functions. This will assist in improving SCM performance, enhancing business performance and ultimately leading to competitive advantage. Ongoing work includes investigating and evaluating the impact of IOS on the supply chain of virtual products and services, as well as the impact of new technologies such as wireless applications.

The companies involved in this study were international or Scottish based companies and therefore the international cultural dimension was not investigated or noted. This research could be expanded to investigate any cultural aspects of the power aspect within business relationships, furthering the work of Hofstede (1983) who recognised cultural differences in the power variable.

References

Akkerman, H., Bogerd P., Yucesan E. & van Wassenhove L. (2003) The impact of ERP on supply chain management: Exploratory findings from a European Delphi study. European Journal of Operational Research. Elsevier. Vol 146, Issue 2, pp 284–301.

Ayers J. (2002) Making Supply Chain Management Work: Design, Implementation, Partnerships, Technology and Profits. Auerbach. pp 237–252. Bandyo-padhyay, N. (2002) e-commerce, Context, Concepts and Consequences, McGraw Hill International (UK) Ltd. New York pp 78, 106–108. Barua A., Konana P., Whinston A. & Yin F. (2001) Driving e-business excellence. MIT Sloan Management Review. MIT Sloan School of Management. Cambridge MA. Vol 43, Part 1, pp 36–45.

Benjamin R., De Long D., Scott Morton M. (1990) Electronic Data Interchange: How much competitive advantage? Long Range Planning Journal. Pergamon Press, UK. Vol 23, Issue 1, pp 29–40.

Bensaou M. & Venkatramen N. (1995) Configurations of Interorganisational Relationships: A Comparison between US and Japanese Automakers. Management Science. Institute for Operations Research and the Management Sciences. Vol 41, Issue 9, Sept 1995. pp 1471– 1492.

Boddy D., Boonstra A. & Kennedy G. (2002) Managing Information Systems. Pearson Education Ltd. Essex. pp 167–183.

Bowersox D. & Closs D. (1996) Logistical Management. McGraw Hill International Editions. Singapore. p17.

Christopher M. & Juttner U. (1998) Developing strategic partnerships in the supply chain. Proceedings of the 2nd Worldwide Research Symposium on Purchasing and Supply Management, April 1–3, 1998. Ipsera and Crisps, London. pp 88–107

Clark T. & Stoddard D. (1996) Inter-organisational business process redesign: Merging technological and process innovation. Journal of Management Information Systems. ME Sharpe Inc. Armonk. Vol 13, Issue 2, pp 9–28.

Clemons E. & Row M. (1993) Limits to interfirm coordination through information technology: Results of a field study in consumer packaged goods distribution.

Journal of Management Information Systems. M E Sharpe Inc. Armonk. Vol 10, Issue 1, pp 73–89.

Crook C. & Kumar R. (1998) Electronic Data Interchange: a multi-industry investigation using grounded theory. Information & Management. Elsevier Science Ltd. B.V. England. No 34, pp 75–89.

EDI data. www. Edi-insider.com in Bocij P., Chaffey D., Greasley A. & Hickie S. (2003) Business Information Systems: Technology, Development and Management for the e-business. Edited by Chaffey D. Financial Times/Prentice Hall. Harlow. pp 117–121

Eisenhardt (1989) Building Theories from Case Study research. Academy of Management review. Vol 32, Issue 4, pp 543–576.

Garcia-Dastugue S.J. & Lambert D.M. (2002) Internet enabled coordination in the supply chain. Industrial Marketing Management. Elsevier Science Ltd. Article in press, uncorrected proof, Sept 2002.

Golden W. & Powell P. (1999) Exploring inter-organisational systems and flexibility in Ireland: a case of two value chains. International Journal of Agile Management Systems. Emerald Group Publishing Ltd. Bradford. Vol 1, Issue 3, pp 169–176.

Grieco Jr. (1989) Why supplier certification? And will it work? Production and Inventory Management Review and APICS News. American Production and Inventory Control Society. USA. Vol 9, Issue 5, pp 38–44.

Grossman M. (2004) The Role of Trust and Collaboration in the Internet-enabled Supply Chain. Journal of American Academy of Business. Cambridge, Holywood. Sept 2004. Vol 5, Issue 1/2, pp 391–396.

Haiwook C. (2001) The Effects of Interorganisational Information systems Infrastructure on Electronic Cooperation: An Investigation of the "move to the Middle." PhD Abstract. Proquest Digital Dissertations. www.lib.umi.com/dissertations, Accessed 13.3.2003

Hayes H. (2002) Outsourcing Xbox manufacturing: Microsoft shows the way for successful outsourcing relationships. Pharmaceutical Technology North America. Duluth. Vol 26, Issue 11, pp 88–92.

Heikkila J. (2002) From supply to demand chain management: efficiency and customer satisfaction. Journal of Operations Management. Elsevier Science Ltd. B.V. Vol 20, Issue 6, pp 747–767.

Henderson J. & Venkatramen N. (1993) Strategic Alignment: Leveraging Information technology for Transforming Organisations. IBM Systems Journal. IBM. March 1993. Vol 32, Issue 1, pp 4–16.

Hill C. & Scudder G. (2002) The use of electronic data interchange for supply chain coordination in the food industry. Journal of Operations Management. Elsevier Science B.V. Vol 20, Issue 4, pp 375–387.

Hofstede G. (1983) The Cultural Relativity of Organisational Practices & theories. Journal of Intenrational Business Studies. JSTOR Vol 14, Issue 2, pp 75–89

Jenson R. & Johnston I. The Enterprise Resource Planning System as a Strategic Solution in Ayers J. (2002) Making Supply Chain Management Work: Design, Implementation, Partnerships, Technology and Profits. Auerbach. pp 165–194,

Kanter R. (1994) Collaborative Advantage: The Art of Alliance. Harvard Business Review. Harvard Business Review Press.Boston. Vol 72, Issue 4, July-August, pp 96–108. ISSN 0017 8012

Klein, H. K. and Myers, M. D. (1999). A Set of Principles for Conducting and Evaluating Interpretive Field Studies in Information Systems. MIS Quarterly, Vol 23 Issue1, pp 67–94.

Krizner K. (2001) On-time data makes all the difference. Frontline Solutions. Duluth. Vol 2, Issue 5, pp 24–27.

Kwon I. & Suh T. (2004) Factors affecting the Level of Trust and Commitment in Supply Chain Relationships. Journal of Supply Chain Management. Institute for Supply Management. MCB University Press. Tempe. Spring 2004, Vol 40, Issue 2, pp 4–14.

Lancioni R., Smith M. & Oliva T. (2000) The Role of the Internet in Supply Chain Management, Industrial Marketing Management, Elsevier Science Ltd. New York. USA. Vol 29, Issue 1, pp 45–56.

Leavitt, H. J. (1965). Applied organizational change in industry: Structural, technological, and humanistic approaches.

Li F & Williams H. Interorganisational systems to support strategic collaboration between firms in Barnes S & Hunt B (2001) E-Commerce & V-Business: Business Models for Global Success. Butterworth Heineman. Oxford. pp 153–170.

McKeown P (2003) Information Technology & The Networked Economy. Thomson Course Technology. pp 77–79.

Messmer E. (2000) Online supply chains creating buzz. Network World. Framingham Vol 17, Issue 17, pp 146–149.

Mische M. (1992) EDI in the EC: Easier said than done. The Journal of European Business. Faulker & Gray. New York. Vol 4, Issue 2, pp 19–22.

Morris D., Tasliyan M. & Wood G. (2003) The social and organisational consequences of the implementation of electronic data interchange systems; reinforcing existing relations or a contested domain? Organisation Studies, Berlin,

European Group of Organisation Studies. Walter de Gruyter & co. Germany. Vol 24, Issue 4, pp 557—574.

Papazoglou M. & Ribbers P. (2006) e-Business: Organisational and technical Foundations. John Wiley & Sons Ltd. England. pp 2–49, 53–72, 368–372.

Premkumar G. Interorganisational Systems and Supply Chain Management: An Information Processing Perspective in Ayers J. (2002) Making Supply Chain Management Work: Design, Implementation, Partnerships, Technology and Profits. Auerbach. pp 367–389.

Santema S (2003) E-Business in Dyadic Supply Chain Perspective. Proceedings of the 19th IMP Conference Lugano, Switzerland.

Sherer S. (2005) From supply-chain management to value network advocacy: implications for e-supply chains, Supply Chain Management: An International Journal. Vol 10, Issue 2, pp 77–83.

Shore B. (2001) Information sharing in global supply chain systems , Journal of Global Information Technology Management. Ivy League Publishing. Marietta. Vol 4, Issue 3, pp 27–50.

Soliman K. & Janz B. (2003) An exploratory study to identify the critical factors affecting the decision to establish Internet-based interorganisational information systems. Information & Management, Elsevier. BV Amsterdam. Vol 41, Issue 6, pp 697–706. ISSN 0378 7206

Stephens S., Gustin C. & Ayers J. Reengineering the Supply Chain: The Next Hurdle in Ayers J. (2002) Making Supply Chain Management Work: Design, Implementation, Partnerships, Technology and Profits. Auerbach. pp 359–366.

Sweet P. (2003) Slowly weaving supply chain webs. Conspectus.com. www.conspectus.com/2003/july/article1.asp. Last accessed February 2006

Venkatramen N. (1991) IT Induced business reconfiguration in M.S. Scott Morton (ed) The Corporation of the 1990s: Information technology and organisational transformation. Oxford University Press. New York in Papazoglou M.P. & Ribbers P.M.A. (2006) e-Business: Organisational and technical Foundations. John Wiley & Sons Ltd. England. pp 60–70.

Venkatramen N. (1994) IT Enabled Business Transformation: From automation to business scope redefinition. Sloan Management Review. Winter 1994 in Papazoglou M.P. & Ribbers P.M.A. (2006) e-Business: Organisational and technical Foundations. John Wiley & Sons Ltd. England. pp 60–70.

Vlosky R., Wilson D. & Vlosky R. (1997) Closing the interorganisational information systems relationship satisfaction gap. Journal of Marketing Practice: Applied Marketing Science. MCB University Press. Vol 3, Issue 2, pp 75–86.

Walsham G. (1995) Interpretive Case Studies in IS Research: nature and method. European Journal of Information Systems. MacMillan Press Ltd. Vol 4, Issue 2, pp 74–81. ISSN 0960 085X

Williams M. & Frolick M. The Evolution of EDI for Competitive Advantage: The FedEx Case in Ayers J. (2002) Making Supply Chain Management Work: Design, Implementation, Partnerships, Technology and Profits. Auerbach. pp 185–194. ISBN 0 8493 1273 6

Wilson M., Stone M. & Woodcock N. (1996) Managing the Change from Marketing Planning to Customer Relationship Management. Long Range Planning. Pergamon. GB. Vol 29, Issue 5, pp 675–683.

Zank G. & Vokurka R. (2003) The Internet: Motivations, deterrents and impact on supply chain relationships. S.A.M. Advanced Management Journal. Society for Advancement of Management. Cincinnati. Spring 2003. Vol 68, Issue 2, pp 33–45.

Rational Exuberance and Revival of the U.S. Automotive Sector

Balkrishna C. Rao

ABSTRACT

The recent woes of the US manufacturing sector have prompted the search for an effective solution for its resuscitation. Such a plan can aid the big three automotive companies in regaining their market share. This article takes a holistic approach in addressing this issue and attempts to, at least, provide a new perspective on the problem.

Comeback of Manufacturing

In this era of globalization, the robust U.S. economy affects the economic workings of the rest of the world in one way or another. With the dot com boom a relic of the past and the real-estate boom showing signs of flagging, this country will need a boom in another area to galvanize consumer spending. In this vein, the feasibility of inducing the next boom in the manufacturing sector should be mulled over by the U.S. government. The manufacturing sector referred to in this article encompasses the quintessential aerospace and automotive divisions together with all other industries, such as pharmaceuticals, semiconductors and computers that use manufacturing operations.

This sector has been the backbone of the American industrial revolution in the past and has greatly contributed to this country's superpower status. It accounted for about 14 % of the U.S. GDP and 11 % of total U.S. employment as of 2003 down from 22 % of total employment in 1977 (US Department of Commerce, 2004). The global lead maintained in the past by the U.S. manufacturing sector has resulted in a pool of workers with enhanced productivity. In fact, the labor productivity in manufacturing doubled during the 1977-2003 period (US Department of Commerce, 2004). Therefore, considering the excellent infrastructure and manpower residing in this sector, it might be possible to tap into its potential for a bright future. China's graying population is another incentive because manufacturing will make its comeback in America with China facing labor shortage in the future. Manufacturing is also a giant in terms of inciting immense activity in a wide variety of other areas that span from raw materials right down to healthcare and finance (US Department of Commerce, 2004). And this boom will be distributed across the width and breadth of America. And as we proceed with the 21st century, progress in this sector will be of import because of its application in many scientific endeavors and production of various finished products. Moreover, the size of the U.S. manufacturing sector makes it an apt candidate for a pilot program to test the efficacy of a subsidized national pension and healthcare plan for improving both its work culture and intrinsic value. Considering its past profitable performance, a properly resuscitated manufacturing sector should be able to account for about 20 % of U.S. GDP which is close to Japanese and German manufacturing's share of total GDP in 2004 (The Economist, April & May 2004). This rebound in the manufacturing sector will boost the U.S. economy and a strong U.S. economy will have positive ramifications for economies around the world. This success could pave the way for a larger national pension and healthcare system built on the earnings of a thriving manufacturing sector. This larger system could in the long run cover all sectors of the U.S. business enterprise inclusive of manufacturing.

Is it possible to stimulate the manufacturing sector with the U.S. already reigning the realm of services industry? Much of the manufacturing sector is sagging under the burden of health-care and pension outlays they owe to their current and retired work force. If the U.S. government could undertake these healthcare and pension liabilities then the U.S. manufacturing sector could compete with its

Asian and European counterparts on an even keel. The absence of this financial burden will help manufacturing companies to enhance innovation for getting involved in cutting-edge technologies for coping with the looming energy and environmental crises and that in turn will aid in counterbalancing the run-of-the-mill manufacturing operations outsourced to Asian countries. Freedom from these stupendous liabilities will allow manufacturing companies, with emphasis on the automotive division, to concentrate on another serious issue "quality" which has bedeviled entrepreneurs since the Japanese forayed into the manufacturing sector.

According to a research effort published by McKinsey & Company (Johnson, 2005), globalization is prodding businesses in the developed countries to encourage workers with innovative skills. With routine chores being outsourced to emerging economies, companies, including those in the manufacturing sector, need to nourish "tacit" (innovative) workers and encourage the transformation of low-skilled laborers into the innovative realm. The encouragement of tacit workers and special training programs for the low-skilled workforce can be achieved in the manufacturing sector with ease by relieving the industrial firms of their legacy costs. The special training programs developed by the manufacturing firms for elevating the low-skilled workforce will do good to the community in general allowing these companies to score in terms of a social context. In this regard, the Ford Motor Company's recent initiative to adopt innovation is noteworthy considering the financial woes troubling the automotive sector. By embracing innovation, the intrinsic value of any commercial setup (for all sectors inclusive of manufacturing) is augmented and this can enhance share value without taking recourse to mechanical schemes such as share buy-backs. A model for innovation that could be followed by the manufacturing sector is that of 3M. It has been very successful in developing, and continues to develop, a wide array of new products by encouraging its workforce to think outside the box. Even Google has adopted this initiative by allowing its workforce to use their imagination, for a certain portion of their official working time, to pursue new ideas. In this era of globalization, prescient adoption of disruptive technologies in developed economies such as the U.S. would neatly balance the routine tasks outsourced to a global workforce in the rest of the world. Besides embracing disruptive engineering technologies, the manufacturing sector should integrate itself with information technology which is revolutionizing the global business enterprise and will continue to do so in the future. Innovation-based initiatives should be encouraged in relevant U.S. business sectors, and not just manufacturing, considering the ever increasing global work force available in the rest of the world for varied tasks.

Over and above the shoring up of manufacturing efficiency through innovation and quality-consciousness, the absence of legacy costs can facilitate the sector in expanding its work force and also improve wages. This underwriting of the massive expenditure by the cash-strapped U.S. government might be termed as "rational exuberance." But this exuberance might just jump-start an ailing, but important, sector of the U.S. business enterprise that will keep the economy rolling.

Bequeath Legacy-Costs?

The first step in achieving this exuberance would be the raising of a staggering amount of seed money for under-writing the pension and health-care costs. Unlike countries such as natural-resource-rich Australia and oil-rich Norway that can channel their budget surpluses from an almost single source into national pension funds, the U.S. government will have to consider various sources of capital. The capital squeezed out of the disparate sources could be put together to raise the huge sum. The lines that follow describe some possible disparate sources.

One source could be a suitable portion of the overwhelming foreign investments flowing into U.S. debt securities. If U.S. treasury bonds and notes do not appear appealing then new bonds, called manufacturing bonds, could be issued by the government for this cause or a combination of these new bonds and traditional U.S. treasuries could be used for raising the capital. Another source for this initial capital could also be the corporate and personal income tax revenues collected by the U.S. government. Yet another source could be based on the Australian scheme with the U.S. government indulging in trading commodities needed by hungry emerging economies in South Asia and South America. Besides trading natural resources, the U.S. government could also look into trading technology with these countries. The U.S. private sector, another potential source, could be enlisted for help in this endeavor. There are certain companies in the manufacturing sector whose profitability has progressed at a decent clip. Examples of such companies exist in the computer and semi-conductor industries, where manufacturing prowess are harnessed to fabricate microprocessor chips and related hardware. This despite the tremendous outsourcing experienced by semiconductor industries to keep themselves profitable. Recently, industrial tycoons such as Bill Gates and Warren Buffet have unleashed their valuable business acumen and gobs of money into global philanthropy. Their philanthropic organizations could be beseeched to contribute generously for this cause. Oil companies could be entreated to generously funnel some of their high profits for this noble cause. The aerospace industry is another profitable group that could be encouraged to help its manufacturing brethren at this time of critical need. A source already existing would be the stock of capital invested in government organizations such as the Social Security Trust Funds and Medicare for the present manufacturing workforce which could be pooled together for this task. Of course, the U.S. government would have to stand guaranty for the anxious workforce whose active funds would be transferred. Any attempt to raise this staggering amount of seed money will also require the U.S. government to redeem locked sources of useful capital. Certain non-performing government assets could be liquidated for raising a portion of the initial capital. Even symbiosis with developed countries like Canada could be considered for this purpose.

The debt incurred through a combination of the disparate sources mentioned above could be honored at a suitable period in time by taxing the then thriving manufacturing sector. Tax rates on both corporate revenue and workforce income could be increased for a suitable period of time to level this debt. These increased tax rates could be phased out at a time when the debt is relieved. An alternative to increasing the tax rates could be a value-added-tax (VAT) wherein a small tax could be added at those stages of fabrication where manufacturing processes are employed. The VAT could be applied to a wide swath of manufactured products from industries that encompass semiconductors, computers, pharmaceuticals, automotive vehicles, aerospace etc. Here again the VAT should be discontinued after leveling the debt. Beyond all these redemption plans, there is always the possibility that the new manufacturing sector armed with innovative techniques and products and quality-consciousness will reap rich rewards for the U.S. economy by exporting its wares to the rest of the world.

The important task of raising the massive amount of seed money could be achieved through a non-profit federal body created for this purpose. This body should also be authorized for financial activities to maintain a substantial stock of capital for future needs through investment options that it deems fit. These options could include hedging strategies, commodities-trading, investing in hedge funds, equity, investments in emerging economies and investments in OECD countries, to name a few, to maximize the returns over long-term. Even investments in this country in booming non-manufacturing sectors could be employed to maintain the cash flow. Such a federal body could also look at the investment strategies employed by some government pension funds, e.g. Singapore, Australia, Norway and even CalPERS, the pension and healthcare system of the state of California, to subsidize wholly or partially the legacy costs of their workforce. If properly invested, the returns in the long-term could be used by this federal body to negate legacy costs for the U.S. business enterprise at large. The second stage of the exuberance endeavor would involve the funneling of this seed capital to traditional government organizations such as the Pension Benefit Guaranty Corporation (PBGC), Medicare and Medicaid for covering the pension and healthcare liabilities of the manufacturing workforce. These organizations, which have existed for a substantial period of time, possess the non-financial wherewithal necessary for dispensing the capital to the companies comprising the manufacturing sector. Their use would minimize the costs associated with the tedious task of directing the cash flow to the individual companies. The roles of these organizations might have to be expanded in this new context to accommodate member companies which are not receiving their benefits. The Pension Benefit Guaranty Corporation (PBGC), which currently offers pension protection to numerous Single-Employer and Multiemployer based companies, incurred a loss of about $ 23 billion for 2005 (PBGC, 2005). Despite its loss-making status, the PBGC could be used only as a means for capital dispensation and with time, maybe, a thriving manufacturing sector might haul it out of the red. Similarly, expenditure for healthcare should be managed through the dispensation systems already in place for Medicare and Medicaid. The ideas outlined in this article are just some of the myriad alternatives to save a critical sector in poor health. A thorough analysis of all the sources is beyond the scope of this article.

Jump Starting the U.S. Automotive Sector

The automotive and aerospace divisions have been two of the major representatives of the U.S. manufacturing sector with the latter being well-heeled. The automotive bloc has accounted for about 1.6 % of the total U.S. GDP in 2003 (US Department of Commerce-II, 2004). Therefore from an economic viewpoint it stands to gain from a revamp. For its revival, the advent of rational exuberance should be followed by the U.S. automotive division infusing "quality," in addition to innovation, into its cherry-picked brands. Improving the quality of a few selected automotive brands will make this division competitive and, maybe, even supplant the products coming from their foreign counterparts. This notwithstanding a beneficial dollar exchange-rate that makes Japanese cars cheaper. The case for fewer brands is corroborated by the sweeping success of the Japanese carmaker Toyota which has 4 brands as opposed to 15 and 8 by General Motors and Ford respectively (The Economist, Sept. 2005). This large variety of automotive vehicles manufactured by the American big three has crimped their bottom lines to some extent. Another example for limited brands from the U.S. manufacturing sector is the aerospace company, Boeing. It has worked on a piecemeal basis on its limited fleet of commercial airplanes which has contributed to its strong position in the business of commercial aviation. This notwithstanding the fact that there are just two global companies catering to the world's need for commercial aviation. With the looming energy crisis, quality will become a significant issue affecting fuel efficiency. Moreover, innovative technologies such as alternative fuel systems will be dependent on quality for these products to compete head-on with their foreign counterparts. This is because high quality automotive vehicles using standard or innovative technologies will consume lesser fuel and require lesser maintenance. Innovation without quality will result in novel U.S. technologies being put to extremely good use in the hands of a foreign competitor who has a workforce that believes in quality. In this regard, as mentioned previously, U.S. automotive companies will have to encourage their tacit workforce and develop training programs for the lower-skilled ones. But the entire workforce, both tacit and low-skilled, will have to be encouraged to be quality-conscious. The latest wave of car models churned out by the American big three have excelled in the aesthetics of appearance. It should be easier to cherry-pick a limited number of these brands that have been well received by customers and continue working on them with innovation- and quality-based efforts to fortify their market share. In such a scenario, a U.S. automotive company's limited brands, with some using innovative alternative fuel technologies, would work so well that both domestic and foreign customers would help them gain top rankings in terms of global production and operating margins. A flourishing automotive division will in turn spawn innovative business opportunities for automotive parts makers such as Delphi. The combination of government's-rational-exuberance, innovation (new fuel systems and other novel technologies), fewer brands and quality can aid in the timely revival of the U.S. auto industry to capitalize on the demand foreseen both in the U.S. and in emerging economies such as China and India.

References

Economic Structure: Japan (April 2004). The Economist.

Economic Structure: Germany (May 2004). The Economist.

Extinction of the Predator (September 2005). The Economist.

Johnson BC, Manyika JM and Yee LA (2005). The Next Revolution in Interactions. The McKinsey Quarterly. 4.

Pension Benefit Guaranty Corporation (2005). Performance and Accountability Report, Washington, D.C.

U.S. Department of Commerce-I (2004). Manufacturing in America: A Comprehensive Strategy to Address the Challenges to U.S. Manufacturers, Washington, D.C.

U.S. Department of Commerce-II (2004). U.S. Automotive Parts Industry Assessment, Washington, D.C.

A Neuroanatomical Approach to Exploring Organizational Performance

David Gillingwater and Thomas H. Gillingwater

ABSTRACT

Insights gained from studying the human brain have begun to open up promising new areas of research in the behavioural and social sciences. Neuroscience-based principles have been incorporated into areas such as business management, economics and marketing, leading to the development of artificial neural networks, neuroeconomics, neuromarketing and, most recently, organizational cognitive neuroscience. Similarly, the brain has been used as a powerful metaphor for thinking about and analysing the nature of organizations. However, no existing approach to organizational analysis has taken advantage of contemporary neuroanatomical principles, thereby missing the opportunity to translate core neuroanatomical knowledge into other, non-related areas of research. In this essentially conceptual paper, we propose several ways in which neuroanatomical approaches could be used to enhance organizational theory, practice and research. We suggest that truly interdisciplinary and collaborative research between neuroanatomists and organizational analysts is likely to provide novel approaches to exploring and improving organizational performance.

Keywords: neuroanatomy, organizations, organizational performance, interdisciplinary research

Introduction

Organizational analysis—the study of organizations as entities within an environment, their social structures, technologies, cultures, physical structures and processes (Hatch & Cunliffe, 2006)—is one of the most complex yet important areas of business and management research. Many different approaches have been employed to assist our understanding, and improvement, of organizations and their performance. Several of these have used comparisons with other entities or explanations—often in the form of a metaphor or analogy—as a means to try and convey images or concepts associated with an organization or organizations.

One of the most significant recent advances in our understanding of organizations and organizational performance has stemmed from comparing organizations to biological organisms. As Morgan (2006) puts it, we find ourselves thinking about organizations as artificial "living systems," existing in a wider environment on which they depend for the satisfaction of various needs. This awareness has led to a focus on understanding the principle that organizations, like organisms, are "open" to their environment. Thus, organizations need to be able to monitor, as well as respond appropriately and rapidly to, changes in their internal and external environments to maintain homeostasis (their primary "physiological" state). Imbalance resulting from an inability either to sense or respond to changes in internal and/or external environments can be potentially disastrous for an organization—resulting in "strategic inertia" (the inability to generate commitment), "strategic dilution" (the inability to provide leadership) and/or "strategic drift" (the inability to focus on where the organization is heading) (Freedman, 2003). The detrimental consequences of "strategic drift" have been identified in companies experiencing rapid change (c.f., Pauwels & Matthyssens, 2003) and also in some extreme cases across an entire industrial sector (e.g. the US auto industry, including such well-known companies as General Motors and Ford of America (c.f., Womack et al., 2007)). Individual organizations with enhanced sensing capacities and capable of adapting themselves to best fit their current and—more importantly—future operating environments are therefore likely to be those that we refer to as high-performing organizations, who will be "superior," gaining competitive advantage over their rivals and in the process survive and prosper (c.f., McKnight et al., 2001; Christensen et al., 1998; Venkatraman & Ramanujam, 1986; Hall, 1980).

These fundamental insights have led some organizational analysts to consider the human brain as a useful metaphor for thinking about the nature and performance of organizations. In 1972 Stafford Beer published what in many respects can be regarded as the first serious attempt to devise a coherent if eclectic model of organizational control (Beer, 1972). Based on physiological principles derived from studies of neuronal function and dissatisfaction with synthetic electronic brain models (such as Ross Ashby's "Homeostat" and Grey Walter's "Tortoise") (Asaro, 2006), his concern lay primarily with developing and extending the application of cybernetics into the realm of management through the derivation of a systemic model of control—eventually to take the form of the highly influential "Viable System Model" (VSM) (Christopher, 2007). More recently, the brain metaphor has been used to draw attention to the importance of information processing, learning and intelligence, providing a frame of reference for understanding and assessing organizations (Morgan, 2006). In terms of organizational performance, this metaphor-based prognosis suggests that organizations should foster moves toward "holographic organization"—as team-based and client-centred self-organized learning systems that place primary emphasis on being open to enquiry and self-criticism (c.f., Nonaka & Takuchi, 1995). Other uses of the brain metaphor tend to be more partial, for example, in assisting with the generation of flexible production system models (Garud & Kotha, 1994).

The brain-based metaphor has also been employed with respect to "softer" aspects of neuroscience, such as cognitive mapping (Tegarden & Sheetz, 2003) and the notion of contemporary organizations as so-called "intellectually impaired bureaucracies" (Ambrose, 1995). Similarly, the notion of memory as applied to organizations ("organizational memory" or "organizational remembering") has taken advantage of brain-based metaphors, leading to the use of terms such as "the brain-based organization" (Walsh & Ungson, 1991; Harari, 1994; Feldman & Feldman, 2006). Thus, the potential benefits that could result from awareness of, and adoption of biological principles adapted from, such brain analogies are well recognised: "Today, knowledge—or more colloquially, brainpower and intelligence—has become the key determinant for economic and business success" (Harari, 1994).

That brain-based approaches can be valuable beyond their use as a metaphor can be found from other fields of research. For example, neuroscience-based principles have been successfully adopted in order to develop a new field in microeconomics—"neuroeconomics." As Camerer et al. (2004) put it, neuroeconomics uses knowledge about brain mechanisms to inform economic theory by opening up the "black box" of the brain, much as organizational economics opened up the theory of the firm with the introduction of organizational behaviour.

The principal concern of this approach is with bettering our understanding of individuals' preferences and preference-seeking strategies and their decision-making processes (McCabe, 2003). More contentious developments have seen the application of functional neuroimaging to market research and marketing science (what has come to be called "neuromarketing" (Lee et al., 2007)) and the application of neurobiology to gambling addiction (the so-called "dopamine hypothesis" (Melis et al., 2005; Vrecko, 2008)).

The most recent development of relevance to our paper has seen the emerging branch of cognitive neuroscience that brings together social psychology with neuroscience—social cognitive neuroscience—and its application to organizations (Senior & Butler, 2007). At its core is an understanding of the relationship between the brain and social interaction, in this case to the study of human behaviour in, and in response to, organizations—what has been termed "organizational cognitive neuroscience" (Butler & Senior, 2007). Here the focus is on the neuroscience of social interactions at the social level and the need to understand the motivational and other social factors that drive behaviour and experience in the real organizational world rather than on organizational performance per se.

In this paper, we propose a new theory-informed and interdisciplinary approach to build on the use of the brain as a metaphor/analogy for understanding the nature and performance of organizations, incorporating long-standing and recent neuroanatomical principles underlying the structure and function of the brain into organizational theory and research. Much of the impetus for this approach has come from the fundamental insight that the anatomical arrangement of the brain shares many of the same constraints and demands that are faced by contemporary organizations, including: costs and efficiency (how much energy provision and blood flow are required?), space requirements (how much space in the skull should the brain really have?) and return on investment (how much could be gained by making the brain 5% larger and/or faster?).

Our goal in writing this paper is to highlight the significant interdisciplinary and collaborative potential that exists for sharing insights and approaches between neuroanatomical research and research on organizational performance. In our view, such a project—what we term "neuroorganizational research"—like that of the development of "organizational cognitive neuroscience" noted above, lies beyond the capacity of either neuroanatomists or organizational analysts to pursue independently since the nature of the problems identified are beyond the scope of either discipline (Klein, 1990). Our hope is that the neuroanatomical principles detailed below—purposefully simplified—will serve to highlight and stimulate new collaborative research avenues for neuroanatomists and organizational researchers alike. Whilst most of the organizational examples referred to relate to service organizations in the transport industries (reflecting the research interests of the co-author), we envisage that neuroanatomical principles could be of potential benefit for organizations of all types and sectors. Moreover, the belief that organizations are individual living organisms that have to be best suited to their local environment to survive and prosper is our fundamental working assumption (c.f. Morgan, 2006)—one set of "rules" may work for one and not for another. Organizations are as individual as each of us, just like our brains.

The rest of the paper is divided into five main sections. In each section we highlight a different neuroanatomical concept or approach that may be of benefit for use in organizational research, including research ideas and possible topics.

Structural Subdivisions in the Brain and in Organizations

In this first section, we outline several existing structural (e.g. anatomical) similarities that link brains and organizations and suggest simple ways in which the adoption of basic neuroanatomical approaches could be used to better define and analyse the performance of organizations.

Many previous studies using the brain as a metaphor for organizational purposes have tended to take a simplistic approach, viewing the brain as a single homogeneous machine, capable of processing large amounts of information and undertaking many complex tasks simultaneously. However, detailed neuroanatomical investigation of the brain reveals a rather different entity, characterised by discrete structural units, interconnected via an astonishingly complex yet logical series of "cables." The human brain is therefore subdivided into a number of distinct anatomical regions, each defined by their appearance and/or main biological function(s). The main classical structural subdivisions of the brain are shown in Figure 1 and include the cerebral hemispheres (including the cerebral cortex), diencephalon, brainstem, and cerebellum (for a more in-depth overview of basic human neuroanatomy, see Crossman & Neary, 2005).

The neuroanatomical approach to understanding the brain takes these structural building blocks as its basis, from which more in-depth investigations can be undertaken to try and uncover its myriad of functions. In a similar way, contemporary organizations can be viewed as a set of discrete entities or building blocks as they tend to have a series of distinct business units, often distinguishable on a physical and/or functional basis (e.g. sales, R&D, production, head office).

Importantly, however, it is the information that comes from understanding the patterns of network connectivity within and between distinct brain structures that has provided the real contemporary insights into how brain structure influences function and therefore performance. Regions of grey matter—shown in Figure 2—are formed by collections of nerve cells (neurons), more often than not performing a specific function or set of closely related functions. Some of the most notable grey matter accumulations in the cerebral hemispheres include the primary motor cortex and primary somatosensory cortex, involved with the initiation and regulation of muscle contraction and the perception of general sensory stimuli (e.g. discriminative touch) respectively.

A

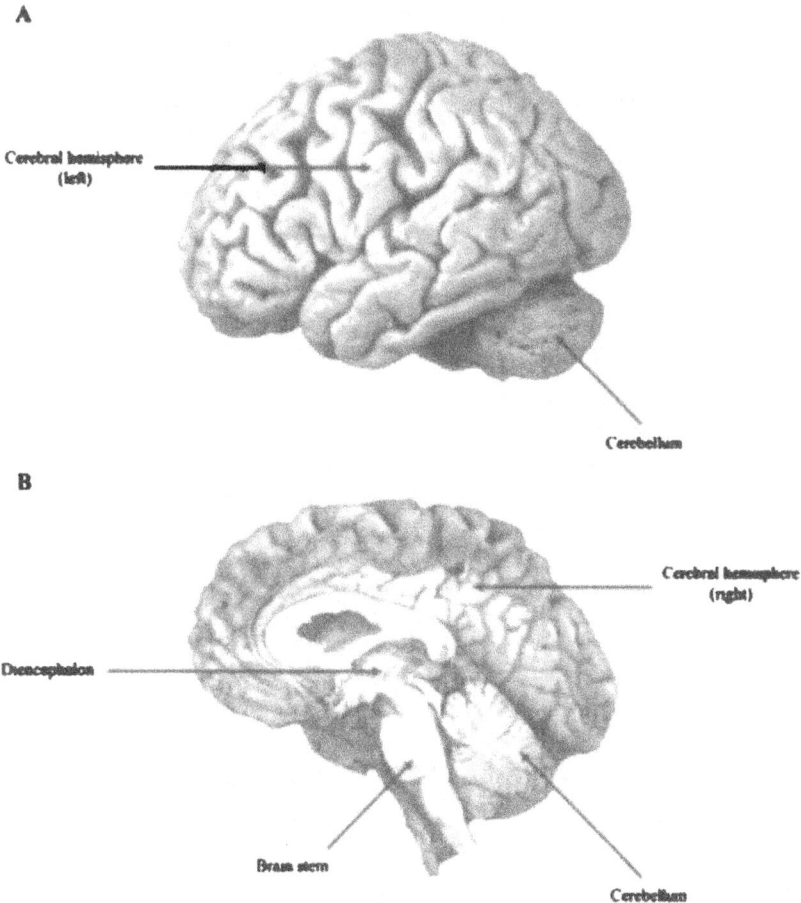

Cerebral hemisphere
(left)

Cerebellum

B

Cerebral hemisphere
(right)

Diencephalon

Brain stem

Cerebellum

Figure 1: Gross anatomy of the human brain and its major constituent parts: A—Lateral (external) view of the left side of a whole human brain annotated to indicate the location of major anatomical subdivisions; B—Medial (internal) view of the right half of a human brain cut in median sagittal section, highlighting other major anatomical subdivisions.

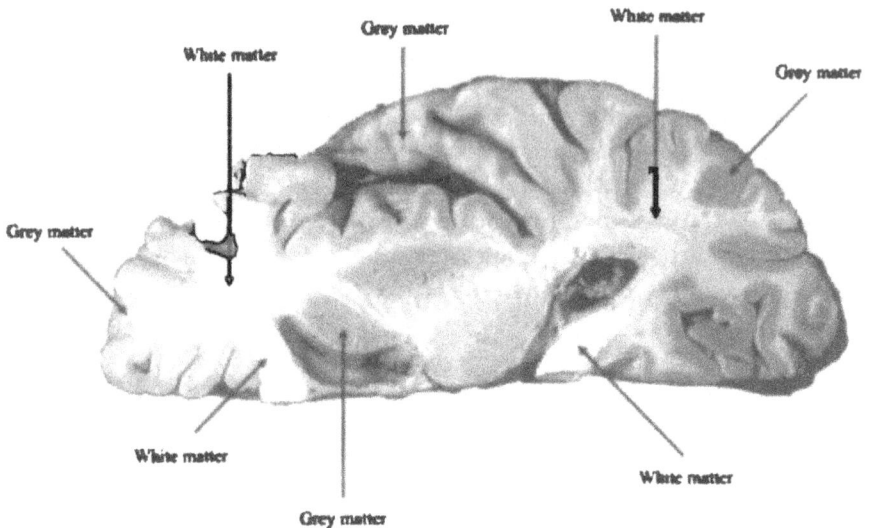

Figure 2: Photograph of horizontal section taken through one half of a human brain. Note the clear distinction between regions of grey matter and regions of white matter.

In the diencephalon, the major grey matter region is the thalamus, a roughly egg-shaped structure sitting beneath each of the large cerebral hemispheres, playing a critical subconscious role in receiving and distributing almost all sensory information coming in from throughout the body. In contrast, white matter regions of the brain form a critical myriad of connections linking grey matter regions to one another (Figure 2). A good organizational equivalence for the role of white matter in the brain, and its relationship to grey matter, is that of internal communication—for example, connectivity between the business units of a global organization using high-speed information and communications technology (ICT). If an organization's offices in New York and London are thought of as being equivalent to two distinct regions of grey matter in the brain, then the streaming of data via fibre optic cables and remote satellite is equivalent to the white matter. White matter therefore provides the all-important communication channels through which different functional units of the brain interact, facilitating rapid (almost in "real time") and accurate transfer of information throughout the entire organ. As a result, the importance of white matter integrity to overall brain function and performance cannot be overemphasised: for example, the specific breakdown of brain regions involved in communication is known to play an important role in debilitating conditions such as stroke and multiple sclerosis (Ferguson et al., 1997; Dewar et al., 1999; Wishart et al., 2006).

The large-scale connections within the brain (white matter tracts) are readily identifiable on gross brain dissections (see Figure 2) and have led to detailed structural maps of the brain (e.g. Wakana et al., 2004). However, recent breakthroughs in neuroanatomical research have allowed researchers—using novel brain imaging techniques (Catani, 2006; Lichtman et al., 2008)—to begin to piece together more subtle, yet no less influential, "wiring" patterns of the brain. This neuroanatomical research has been key to our contemporary understanding of the structure and function of the brain, as mapping refined physical and functional neural circuits has allowed us to account for brain activity and observe how alterations in discrete regions of the circuitry can lead to dysfunction of the brain.

That a similar structural arrangement can be found in organizations becomes more obvious when examining how groups of individuals become ordered within an organization and the routes of communication they use. Many organizational subdivisions are directly equivalent to grey matter nuclei in the brain as they often contain large numbers of individuals (equivalent to neurons in the grey matter), more often than not focused on performing a specific task or role. White matter equivalents include the internet, email, telephone, meetings, video conferencing, etc. As with the brain, the integrity and efficiency of formal and informal intra-organizational communication link s are known to be similarly important for the successful function of high-performing organizations (Adams et al., 1993; Lievens & Moenaert, 2000; Marshall et al., 2007). Thus, a mapping approach akin to "business process mapping" (Kaplan & Norton, 2000), but based on neuroanatomical principles and employing neuroanatomical terminology rather than process modelling methods like "Six Sigma," is likely to provide a sounder basis from which to develop a coherent and relevant understanding of an organization's structure and functions, as well as providing the opportunity to identify regions of potential weakness or dysfunction in its connectivity and performance.

Whilst this approach may seem rather unproblematic, gaining the required information to build a "grey and white matter wiring diagram" for an organization may not be so simple. For example, one of the key concepts underlying connectivity within the brain is that of the segregated nature of information coding, transmission and decoding (Ciborra, 2001a). Communication channels in the brain are almost always divided into those with input functions (sensory pathways carrying information about the internal and external environment towards the brain), modulating functions (short, local networks allowing communication between adjacent regions of grey matter), or output functions (motor pathways carrying effector information away from the brain and out into the body, leading to contraction of muscles). These different modalities can be readily distinguished from one another in the brain as a result of the selective use of different chemical signals (neurotransmitters) released by distinct subpopulations of neurons. In the organizational context, such clarity is less readily observed, although no less important. For example, individual business units require a clear understanding of the location and nature of incoming information versus intra-unit communication channels and outgoing communications. However, the contemporary world of rapid and integrated communications technology and media (e.g. the internet, intranet, wire-less communication) makes deciphering the source and relevance of information a time-consuming and sometimes overwhelming chore, contributing to the phenomenon of "information overload" (Eppler & Mengis, 2004). Whilst it can be argued that such clear subdivisions are rarely the norm in most organizations (Kaplan & Norton, 2000), a good example of where this approach can be identified in organizational practice is provided by the organization of air traffic control services (Smolensky & Stein, 1998)—an intensely information-rich and communications-critical operating environment, where there is an unambiguous, overriding and universally agreed goal (the safe arrival of aircraft). Studies of such organizations and operating environments may therefore provide a good starting point for future research investigating the potential for incorporating neuroanatomical principles into the structure and function of high-performing organizations.

'Doing' Individuals and 'Supporting' Individuals

In this second section, we examine the way in which cellular neuroanatomy (investigating the identity, relationships and function of individual cells within the brain) has provided novel insights into the structure and strategy of the brain and outline potential organizational equivalences.

One of the key principles of cellular neuroanatomy is that the brain consists of two main populations of cells: neurons and supporting cells (glial cells). The electrically excitable neurons are the primary "doing" cells of the brain. Their cell bodies are resident in grey matter whilst their long thin processes (axons) pass out into the white matter in order to establish contact with cells in other regions of the nervous system. These axon processes end at specialised chemical communication points known as "synapses," from where they pass on their information to the next cell in the network (known as the post-synaptic cell). Glial cells, in contrast, assist neurons to improve their performance by providing nourishment, shielding them from the immediate internal environment, and moderating their exposure to potentially damaging chemicals and environments. They are present throughout the grey and white matter, but in the white matter play an important role in ensheathing (myelinating) the axons of neurons to increase the speed at which they conduct electrical impulses via saltatory conduction. Importantly, this very clear delineation of cell type and function is set, and normally cannot be reversed after each cell has undergone development and maturation. Thus, "doing" neurons are specialised precisely for that function and do not have to worry about supporting any other individuals. "Supporting" glial cells are the complete opposite, sustaining other cells wherever they can but without the requirement or ability to directly generate or transmit any information.

The importance of delineating cells with a "doing" function from those with a "support" function suggests that individuals within an organization (akin to cells within the brain) may also perform better when equipped with the specialised skills and resources to perform one of these roles, rather than trying to do both. There are numerous examples where individuals within organizations have attempted to take on both roles, leading to work over-load, a lack of specialisation, and the threat of failure. A particularly apt example is provided by the iconic US low-fare, high-value airline JetBlue Airways and its founder, chairman and CEO, David Neeleman. According to one seasoned industry observer (Straus, 2009), the carrier's highly publicised operational and financial turbulence in early 2007 could be attributed in no small part to Neeleman's attempts to both steer the airline's future development (in his role as chairman) as well as maintain operational control (as CEO). Within the space of a few months, JetBlue effectively restructured its entire operation. A JetBlue spokesperson told Straus that "The big difference now is that David simply could not do both. He could not run the day-to-day and also look three, five, 10 years out into the future. This change allows him to do that. ... Dave [Barger's] strength [as the newly appointed CEO] is running an airline. He knows what it takes to restore JetBlue's operational integrity."

Despite the distinct identity and roles of neurons and glia in the brain, one of the most important breakthroughs in cellular neuroanatomy research has been to highlight the requirement for a close working relationship between the two cell types in order to achieve optimal performance (Sherman & Brophy, 2005). Changes in the status and/or function of one cell type are almost always immediately detected by the other. Breakdown of these mutual relationships results in altered function and anatomy in the nervous system and is thought to be play an important role in several neurodegenerative conditions, including motor neuron disease (Boillee et al., 2006; Nagai et al., 2007). From an organizational perspective, this suggests that whilst individuals with "doing" or "supporting" functions may benefit from being specialised towards one of these roles, they should never isolate themselves from individuals in the same work team or organization that perform the converse role. Whilst it may be tempting to view "doing" individuals as the most important in any organization, and therefore in some way superior to "supporting" individuals, the cellular neuroanatomy of the brain suggests that mutual respect and support between the two types of role is likely to be important for optimum performance. It is therefore possible to envisage a neuroanatomical approach being beneficial to understanding the identity and roles of an individual within an organization by coding them as primarily a "doing" or a "supporting" individual. The awareness of this identity could then be used as a tool for managing workload and the type of work for any given individual, and could also be used to ensure good communication links between doing individuals and supporting individuals.

These neuroanatomical insights also lend support to the notion that, on the one hand, bureaucratic command-and-control type structures are less than ideal and should be avoided wherever possible, with small interdependent organizational units being preferred, and, on the other, recognizing that all personnel should be integrated into the workings of the organization with a clearly defined role where they are trusted, respected and inspired. Two organizations that it could be claimed currently integrate the symbiotic "doing" and "supporting" functions in a highly transparent form are The Walt Disney Company and IKEA. The lionization and reinvention of Disney and the "Disney approach" under the stewardship of Michael D Eisner have been well documented (c.f., Disney Institute, 2001; Capodagli & Jackson, 1999) but not uncritically (c.f., Bryman, 2004); and IKEA, the Scandinavian-based low cost-high quality furniture retailer—the largest in the world with a growing global presence—is beginning to assume a similar iconic status (Kotler, 1999; Edvardsson and Enquist, 2002). Such high-performing organizations may therefore provide good case study material for further investigations into existing and potential future synergies between organizational and neuroanatomical theories and practice.

Dealing with Information Overload

In this third section, we highlight the main neuroanatomical approaches used by the brain in order to overcome one of the major problems shared by brains and organizations alike: how to deal with information overload. Receiving, integrating and highlighting important information collected from both internal and external environments is undoubtedly one of the core functions of the brain. Organizations face a similar complexity and wealth of information available to them, which is only likely to increase over the coming years due to the expansions in, and demands of, ICT. And yet, organizations that are successful are generally considered to be those that have a "bias for action" (Peters & Waterman, 1982): where communications are of the essence and they are best capable of collecting, processing and responding to information from their internal and external environments (Ciborra, 2001b). Organizations therefore face very similar problems to the brain, being required to process important information rapidly, efficiently and effectively, whilst not losing sight of less specific (albeit potentially no less important) changes occurring in their wider internal and external environments. ICT provides the required speed of response in many organizations, but does not guarantee its effectiveness (Eppler & Mengis, 2004). Any strategy that has the potential to improve the abilities to sense, predict and assimilate changes in an operating environment is therefore likely to be advantageous.

The anatomical arrangement of the human nervous system (and here we are deliberately extending beyond the boundaries of the brain to incorporate the spinal cord and peripheral nerves) provides two distinct ways of dealing with this problem: reflex arcs that are capable of eliciting repeatable and rapid responses to important stimuli (e.g. pain or excessive heat); and the ability to deal with information at a conscious or subconscious level, allowing some functions to proceed with the minimum of input or resources from more complex and resource-demanding conscious areas of the brain.

The presence of reflex arcs within the nervous system allows rapid, stereotyped responses to occur following an important, narrowly-defined, non-strategic (i.e. not requiring any long-term planning) sensory stimulus (e.g. rapid lifting of the foot in response to pain resulting from treading on a sharp object). These responses occur locally within the spinal cord, facilitating a stereotyped response over a much shorter timescale than if the sensory information had to travel up to higher "conscious" regions of the brain, before being processed, assimilated and a decision arrived at as to how to proceed. The presence of reflex arcs therefore bypasses the requirement to sift through large volumes of information that are constantly coming in from all types of different sources in order to identify, process and respond to a critical stimulus that would benefit from a rapid response. The reliance on integrated management information systems in the contemporary organization provides the opportunity to implement similar reflex arcs in the organizational context, allowing much more rapid responses to key incoming stimuli without requiring "decisions" to be made prior to a response (c.f., Larsen & Leinsdorff, 1998). However, the incorporation of neuroanatomical principles underlying the reflex arc into such management information systems might improve their performance.

The cellular arrangements of the nervous system that permit a rapid and stereotyped reflex response are based around the existence of discrete populations of individual neurons in the spinal cord that are specifically tasked, positioned and connected to detect pre-defined "important" events coming in from the periphery (e.g. pain) and which also have the "authority" to initiate a response (e.g. movement of the limb away from the painful stimulus) without prior approval from higher centres in the brain. Building such an arrangement into organizational structure and strategy, most likely by incorporating this approach into pre-existing integrated management information systems, may therefore confer similar reflex abilities.

In organizational terms the basis of such an approach may already exist—albeit without the clarity provided by the neuroanatomical framework—corresponding to the practice of pushing responsibility down the line, combining firm central direction with maximum individual autonomy (what has been called "simultaneous loose-tight properties" (Peters & Waterman, 1982)), and reflected in the obsession with delayering (i.e., removing intermediate levels of managerial responsibilities wherever possible). This approach is currently the sine qua non of two of today's most successful business operations—the "production system model" of the Japanese car maker Toyota (Liker, 2004; Dahlgaard-Park & Dahlgaard, 2007) and the "fast aircraft turnaround model" typified by the low-fare high-value US carrier Southwest Airlines. The "Southwest model" (Flouris & Oswald, 2006), now replicated in part by the majority of low-cost airlines around the world, is based not only on high asset utilisation (aircraft operations) but more importantly on the development of high-performance relationships that characterise the organization (Gittell, 2003). These characteristics suggest that the devolution of decision making—in response to a specific and defined set of environmental stimuli—down to the level of individual workers within an organization is a good way to improve rapid, efficient and effective responses to key, stereotyped stimuli relevant to an individual organization (e.g. competitor price cuts, supply chain problems, growth in customer demand).

For information and inputs that require more processing and integration than those which elicit a simple reflex response, the brain resists information overload by channelling information accurately into discrete anatomical regions operating at conscious or subconscious levels. The ability to channel information into regions of the brain that do not normally require conscious control or awareness allows many core survival functions required for life support to take place with a minimum of effort. For example, subconscious regions of the brain such as the brainstem are constantly monitoring and responding to changes in both internal (heart rate, blood pressure, temperature, etc.) and external environments. Only when the magnitude of change in any of the monitored parameters broaches a critical level does an "awareness" of the information reach an individual's consciousness. This arrangement frees up conscious areas of the brain, such as the cerebral hemispheres, to undertake more complex information assimilation, decision-making and planning activities. Interestingly, it is these conscious areas of the human brain that have undergone massive evolutionary expansion, allowing the development of our higher cortical functions (Kriegstein et al., 2006).

From an organizational perspective, the potential benefits of subdividing information processing and decision-making tasks into "conscious" and "subconscious" areas appear to be worthy of serious further investigation. Examples of where this kind of approach has been shown to work effectively include high-performing organizations like Toyota and Ikea that combine the attributes of a simple structural form and a "lean" management style with firm central direction and maximum individual autonomy (Stadler, 2007)—but, in this case, with a focus not only on developing autonomy among individuals but also on their internal entrepreneurship and innovation capabilities. This approach highlights one potential way to allow such autonomy and individual innovation within an organization, without a loss of control or focus: an obsession with the customer. Nowhere is this better illustrated than in an industry-leading service-driven company like Southwest Airlines, structured on the basis of high-performance relationships that are built on "relational coordination" (Gittell, 2003). For example, Southwest's Customer Service Commitment states that: "We tell our Employees we are in the Customer Service business—we just happen to provide airline transportation" (SWA, 2009). Southwest's remarkable success can be measured by the fact that they are the only airline ever to have recorded year-on-year profitability since their inception in 1971 and to have achieved the best cumulative consumer satisfaction record of any US airline. This has been achieved with low staff turnover rates, more flexible work arrangements than competitors and a high level of unionization (Gittell, 2003; Flouris & Oswald, 2006).

The use of a neuroanatomical approach may therefore be of benefit for assisting other organizations to conceptualise and implement similar structures and practices. Here, core housekeeping operations equivalent to "subconscious" functions should be able to progress with a minimum of disruption and effort, without reliance on constant input from "higher" ("conscious") centres of the organization. Similarly, individuals and managers connected with strategy and planning would not expend time and energy on "lower-level" issues, unless something went wrong and required intervention. However, neuroanatomical research has highlighted that this approach—separating the "conscious" and "subconscious" areas of an organization—does not imply that basic "subconscious" business processes should not be carefully monitored and remain within the conscious psyche of the organization as a whole. Quite the opposite, as failure of "subconscious" business processes often means failure of the organization. Rather, "subconscious" business processes should be allowed to proceed without repeated interventions that rarely add significant value to the organization as a whole.

One likely consequence of adopting such an approach for an organization would be problems with directing relevant information towards conscious or subconscious centres. Once again, however, the brain has already developed a means to deal with this problem that could readily be transferred to the organizational context. The thalamus (the grey matter region located in the diencephalon—Figure 1), sitting beneath each of the large cerebral hemispheres, plays a critical subconscious role in receiving and distributing almost all sensory information coming in from throughout the body (Schmahmann, 2003). One useful analogy is to compare the thalamus to a central postal sorting office, where all mail consignments (sensory information) have to be brought before being sorted by address and sent out to their correct final destination. Information coming in from the body is therefore not communicated to the conscious cortex in a haphazard, randomly organized way, thereby reducing information asymmetry and misinformation. The presence of a similar central (but not necessarily centralised) information receiving/distributing centre could be provided to an organization by modifying and enhancing existing ICT infrastructures (see above; Monteiro & Hepso, 2001) or customer relationship management models (CRM) (Wilson et al., 2002), as they are often already established to collate and distribute important internal and external information (e.g. competitors' prices, internal supply chains, direct and indirect costs).

The ability to separate "conscious" and "subconscious" areas is also critical for undertaking the exceptional long-term planning and strategic thinking functions associated with the human brain. As a result of these attributes, the human brain has the ability not only to monitor and respond to the environment, but also to "second guess" it, allowing plans to be made to influence internal and external environments rather than simply being forced to respond to changes in them. This ability is one that is highly sought after by organizations, forming the basis of much strategic planning and re-organization. Its locus is to be found in the twin themes of organizational leadership (especially the powers of visionary and transforming leadership (Bryman, 1992)) and organizational culture (in particular the ability to foster and engender creativity and innovation in the implementation of strategy). The adoption of a neuroanatomical approach to "conscious" and "subconscious" functions may therefore go some way towards conferring this ability. In particular, executive decision-making and planning processes in the human brain take place in regions of the cerebral hemispheres distinct from other areas responsible for conscious awareness of the environment (e.g. primary sensory cortex) and executing decisions (e.g. primary motor cortex), suggesting that the physical and intellectual separation of these functions from the more mundane day-to-day tasks of running an organization would be beneficial. This approach is perhaps another instance where the high-performance relationships that underpin Southwest Airlines' leadership, culture and strategy (Gittell, 2003) contribute to its unique competitive advantage (Flouris & Oswald, 2006).

Importance of Environmental Exposure During Development

In this fourth section, we examine how insights gained from studying the development of the brain might be useful for our understanding of the birth, growth and development of organizations. In particular, we discuss the importance of environmental exposure during the early stages of an organization's life-cycle: creation and growth. The developmental processes that the brain goes through to reach its final adult state are, just like organizations, highly complex and critical for the successful establishment of a fully-functioning organ. The importance of these early stages of development should not be overlooked, especially when they are known to play a significant role in determining the eventual "mature" state and performance of both brains and organizations (Hensch, 2005; Wolpaw, 2006; Van de Ven, 1980; Quinn & Cameron, 1983).

The gross anatomy of the brain is established early on in pre-natal development, much as the plans and "blue-print" for the general structure and strategy of a new organization (or new business unit within an existing organization) are often established well before the organization becomes a physical entity (Quinn & Cameron, 1983). However, the fine network of internal circuitry within the brain is only refined into its experienced, fully-functioning state by a process of gradual sculpting and re-wiring that occurs during critical periods of development after birth. During these periods, excessive and incorrect connections and pathways are pruned away—as shown in Figure 3.

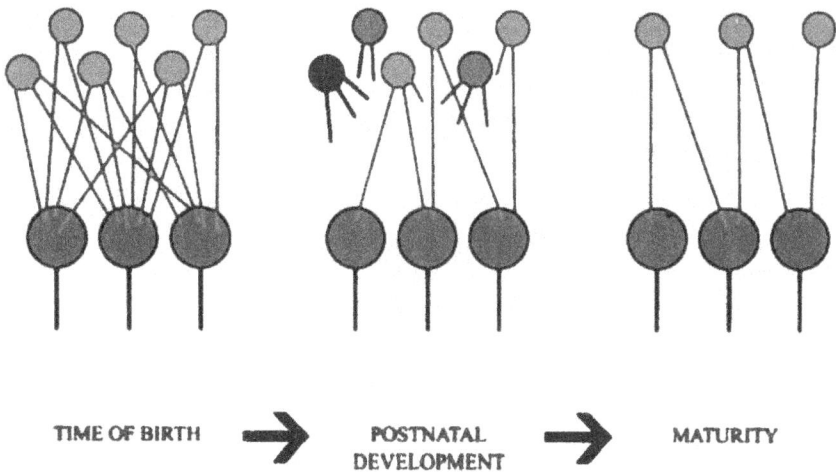

TIME OF BIRTH ➡ POSTNATAL DEVELOPMENT ➡ MATURITY

Figure 3: Schematic diagram representing the normal process of postnatal pruning of connections that occurs throughout the human nervous system shortly after birth.

The removal of excess and/or incorrect connections, critical for establishing the fully-functional mature nervous system, is driven by a variety of genetic and epigenetic factors such as their relative levels of activity, the presence or absence of competing inputs for the same role, relationships of neuronal cells to their surrounding glial cells, and/or suitability for a specific function (Wyatt & Balice-Gordon, 2003; Low & Cheng, 2006; Freeman, 2006). As a result, the brain is prepared for its future tasks and challenges by moulding itself in response to the external and internal environments it becomes exposed to during the first weeks and months of its life.

The importance of environmental exposure is perhaps best illustrated by the effects of visual disturbances on the development of the primary visual cortex (the region responsible for conscious processing and awareness of visual information). It is well established that visual acuity in humans develops substantially (improving roughly fivefold) over the first few months following birth, before fine-tuning the system over the ensuing six years (Maurer & Lewis, 2001). However, if the visual system is perturbed during this critical developmental period (e.g. by congenital blindness or cataracts) then the primary visual cortex fails to re-organise and arrange itself normally, and in some cases can even be taken over by sensory inputs coming from other modalities such as touch or sound (Maurer et al., 2005). As the individual matures, the ability of the brain to undertake re-modelling and reorganization on such a large scale declines. Thus, the majority of experience-dependent learning needs to be undertaken, with accompanying resource allocation, during the early periods of an individual's life, as it cannot be adequately compensated for at a later date.

An understanding of the critical periods present during the successful growth and development of the brain, alongside an appreciation of the consequences of failure to provide environmental exposure during these periods, has potentially important insights for organizational development. Although many different models of the organizational life cycle (OLC) have been proposed (e.g. Lippitt & Schmidt, 1967; Miller & Friesen, 1980; Greiner, 1998), many share the concepts of organizational creation/birth and growth preceding maturity (Lester, 2004). Similarly, many models imply that a failure to undergo developmental processes required to establish a mature organization with a good environmental fit and the capacity to respond to important sensory stimuli results in the high numbers of companies that fail to last beyond 12 to 18 months from the time of their creation (Quinn & Cameron, 1983).

Of particular note, the high levels of resources available to the developing brain after birth (e.g. energy, number of contributing cells) have been found to be of critical importance, allowing the initial (pre-natal) over-elaboration of contacts and functions before subsequent post-natal sculpting occurs. Translation of these insights to the organizational perspective suggests that resource allocation during creation and early growth phases of the OLC is likely to be critical to success (Buenstorf & Witt, 2006). Thus, an organization looking to take advantage of neuroanatomical principles would require the ability to commit significant resourcing to ensure that, at its conception, it has more resources than may eventually be required. This is especially the case for "softer" resources like management experience and commitment, time and creative space, as well as the more conventional resources of superior market intelligence, appropriate personnel, innovative products and services, efficient outputs and effective networks of inter-organizational connections (Lester, 2004). These facilitate the all-important freedoms to subsequently undergo "developmental remodelling" in response to both internal and external environmental pressures and opportunities that are unlikely to be fully anticipated or predicted before launch. In this way, an organization has the potential to be more successful in its growth and early maturity phases because it has allowed itself to be shaped by the real pressures of the turbulent business world it inhabits, rather than trying to "shoe-horn" a "one trick pony" or "one size fits all" offering into a predetermined fixed structure, planned before creation, into a business environment where it does not ideally fit.

An illustration of the potential effectiveness of this approach to "developmental remodelling" is provided by the evolution of JetBlue Airways and its "value-based model" (Flouris & Oswald, 2006; Fiorini, 2002). Neeleman, JetBlue's founder, having already experienced the trials and tribulations of starting a new airline—Morris Air Service—which was subsequently sold to Southwest, was then employed by Southwest during which the idea of JetBlue was born. The four factors which appeared crucial were: the prior experience gained when creating a minor airline; the valuable learning gained whilst at Southwest; the ability to attract and adapt mature airline talent to a new corporate concept prior to start-up; and finally, perhaps most crucially, the reflective evaluation of the reasons behind the failed growth strategy of the path-breaking low-cost US carrier PEOPLExpress (Wynbrandt, 2004; Peterson & Glab, 1995)—and a recognition of the resourcing requirements necessary for creating a value-based service: innovation (in the sense of "doing things differently" to competitors), flexibility, speed of response and a sense of intimacy with employees and customers alike.

An apposite organizational illustration of the sort of "developmental remodelling" that the brain undergoes is provided by the European low cost airline easy-Jet. In the early 1990s, its founder, Stelios Haji-Ioannou, saw the potential not only for a European version of the highly successful Southwest Airlines but the opportunity created by the liberalisation of European skies and the challenge that could be made to the high-cost incumbent airlines' dominance in key markets. As Kumar (2004) points out, the creation of easyJet was not simply a case of the right person being in the right place at the right time but, more significantly, the ability of Haji-Ioannou to develop a truly transformational initiative—a strongly marketing-led strategy based not on conventional market segmentation but on strategic segmentation. As both Kumar (2004) and Rae (2001) illustrate, easyJet as an organization, the "easyJet model" (Sull, 1999) and its founder—like Neeleman and JetBlue—have been the subject of and subject to a continuous evolutionary process of refining, sculpting and rewiring since its conception—perhaps a classic case of organizational fluidity and the importance of environmental exposure during development and the resourcing implications that are a prerequisite for success. Today neither easyJet nor JetBlue bear much of a resemblance to their original organizational states (Kuemmerle, 1999). Such organizations may therefore provide ideal sources for case studies further investigating how neuroanatomical concepts regarding environmental exposure during growth and development may confer a competitive advantage in the organizational context.

The Importance of Retaining Flexibility

In this final section, we discuss how research highlighting the importance of flexibility within the mature brain might be useful for our understanding of the impact and importance of one specific facet of modern organizational research: organizational memory.

Whilst there is a considerable volume of literature detailing the existence and importance of organizational memory, and its roles in organizational learning and decision-making (Walsh & Ungson, 1991; Feldman & Feldman, 2006), many mature organizations are characterised by static structures and procedures that are unwilling and/or unable to remodel themselves in response to changing business environments and competitive pressures (Ambrose, 1995; Brown, 2004), resulting in the threat of "strategic drift" or its realisation (Johnson, 1988; Larsen & Leinsdorf, 1998; Ciborra, 2001a). For example, the construct of "network inertia" has been identified by Kim et al. (2006), referring to a persistent organizational resistance to changing interorganizational network ties or difficulties that an organization faces when it attempts to dissolve old relationships and form new network ties. McKnight et al. (2001) refer to the processes of "creative destruction" that can accompany an organization's quest for survival in ultra-competitive markets like telecommunications in the Internet age. Thus, many organizations lack the core creative abilities possessed by the brain in order to break down and/ or establish new connections, thereby conferring adaptability and "learning"— organizational fluidity.

One way around this problem may be forthcoming from the adoption of neuroanatomical principles. Despite losing the ability to undertake large-scale developmental reorganization after the first few years of life (the reason why it becomes a lot harder to, for example, learn a foreign language the older you get), the brain retains the capability to adapt and remodel itself in response to more subtle stimuli, at the level of individual cells and circuits, throughout its entire life. This ability to undergo functional and structural remodelling is critical to undertake complex adaptive functions in the mature brain, such as learning and memory in response to changes in internal and external environments (Trachtenberg et al., 2002; Cooke & Bliss, 2006).

Importantly, it has been found that most of this plasticity occurs at the sites of connection and communication between neurons: the synapse. Individual synaptic connections are capable of being strengthened or weakened as well as created and lost. The basic principle underlying each of these processes is neural activity: the more functional neuronal circuits are utilised, the stronger the connectivity within it becomes and the more stable they become in the long-term. Thus, the ability to make, break, strengthen and weaken connections in the brain is critical for its function. These processes allow an individual to adapt constantly to the pressures and demands of their surroundings, making the most out of opportunities that become available. The requirement for making and breaking network connections within and between organizations in order to be able to adapt and respond to new and emerging opportunities and threats is supported by the findings of Christensen et al. (1998) into survival strategies in fast-changing industries (in this case, technological innovation in the rigid disk drive industry). Their analysis suggests that simple "first-mover advantages" and most of the postulates of the entry timing literature might not hold true in such industries. Although they claim that entry timing still has something to tell us about the success or failure of organizations in these industries, they propose the idea of a "window of learning" as a more accurate way of conceptualizing the importance of entry timing in fast-paced industries. It also strongly suggests that "creatively intelligent" organizations capable of dynamic responses are likely to be those that succeed in uncertain business environments (Ambrose, 1995; Hart & Banbury, 1994; Hart, 1992).

Two organizations that it could be claimed exemplify this plasticity and ability to remodel their network connectivities in a highly creative and intelligent way are Southwest Airlines and Virgin Atlantic Airways. Like Southwest, Virgin has constantly over its 25-year life sought to gain a competitive advantage by differentiating itself from what it regards as complacent and conservative mainstream airlines. This differentiation strategy has been based on what has been called a "value innovation" approach (Kim & Mauborgne, 1999). As Denoyelle & Larreche (1995) have put it, in its customer's eyes, Virgin represents a sense of value for money, quality, innovation, fun and a sense of competitive challenge. This is reflected in its corporate positioning and recruitment policy: "Immaculate service and unrivalled quality are everything to us here at Virgin Atlantic. The high standards and experience of the people we hire has helped us become one of the world's most highly rated airlines. In fact, whether you join us in the air, on the ground or behind the scenes, you'll need to be totally focused on delivering everything our customers have come to expect of us."(VAA, 2009) In practice, over the years, this has required the airline not only to recruit personnel whose beliefs and personality are in alignment with its corporate values (Mitchell, 1999) but also to monitor customer feedback and then use that feedback as a basis for improving the customer experience on a constant and continuous basis. According to Denoyelle & Larreche (1995), Virgin looks for opportunities where they can offer something better, fresher and more valuable by moving into business areas where the customer has traditionally received a poor deal and where the competition is regarded as being complacent; it is proactive and quick to act. "Because we're such a complex and rapidly evolving business, we expect all our people to be adaptable, quick thinking and people focused every day to contribute to our ongoing success."(VAA, 2009) Unlike its principal competitors, Virgin relies heavily on the power of its brand image and supporting infrastructure: the highly visible and personal reputation of its charismatic leader, Richard Branson; a professionalized management style and structure based on empowerment and minimal layers of management; its dynamic network of influential business partners; a small number of directors in what is a privately owned company; no global corporate headquarters; and being one of a family of businesses in the Virgin Group (Denoyelle & Larreche, 1995; Virgin Group, 2009). Combe & Botschen (2004) contend that this illustrates what they call an effective use of "process developmental strategy": a continuous process of "learning by doing" when confronted by dynamic complexity, where obvious interventions produce non-obvious consequences and where the same action can produce dramatically different effects over different time horizons. They argue that the complexity of managing quality in a consumer-led service industry—which after all is what both Virgin and Southwest see themselves in—is further increased when there is continuous change in the external environment due to intense competition and changing customer needs. Flexibility and flexible response—the requirement for making and breaking network connections in order to be able to adapt and respond to new and emerging opportunities and threats—is clearly critical to organizational success.

However, gaining the creative ability to alter network connections and business processes within and between organizations is not as straightforward as it may seem (Grinyer et al., 1989). Attempts to adapt and remodel both organizational structure and strategy—to "turnaround" an ailing company into a high performing organization—are often constrained by embedded (possibly even inherited) routines, rules and procedures, commonly based around technological constraints and/or organizational culture (Kim et al., 2006) or else a failure to close what has been called the "strategy-to-performance gap" (Mankins & Steele, 2005). Such constraints often result in "deadlocked organizations" (c.f., Nicholson et al., 1990) and, in some extreme cases, to a "deadlocked industry" (Bryman et al., 1996). The neuroanatomical principles we have explored thus far may be of considerable utility to those organizations not conforming to the heights of "corporate excellence" since our core neuro-organizational equivalences offer the ability to "diagnose" organizational problems like "strategic drift" (Johnson, 1988) and "intellectually impaired bureaucracies" (Ambrose, 1995). For example, it is worth noting that recent advances in neurodegenerative research have shown that synaptic connections (communication sites between individual nerve cells) are often disrupted at very early stages of disease in the brain (Wishart et al., 2006). This suggests that connectivity between and within affected brain regions in human neurodegenerative disease is affected very early on, before the onset of more major symptoms. Approaches to improving and maintaining communication links within organizations, such as those detailed in earlier sections of this paper, may also provide useful monitoring tools with which to identify and rectify early "symptoms" of malfunction or breakdown in key business processes prior to the onset of a loss of critical flexibility and the ensuing "strategic drift."

Conclusions & Implications for Future Research

In this paper we have proposed several ways in which neuroanatomical approaches could be deployed to enhance our understanding of how organizations function and perform. We suggest that interdisciplinary research between neuroanatomists and organizational analysts is likely to provide novel insights into ways to explore, conceptualize and improve organizational performance. Such an interdisciplinary approach could help identify combinations of high-performing organizational attributes that could lead to the development of a unique and sustainable competitive advantage for a wide variety of organizations in today's increasingly competitive and turbulent world.

As we have indicated in our selective use of examples, many organizations already appear to be doing several of the things we contend neuroanatomy tells us are important for the performance of the brain, suggesting that they are to all intents and purposes already developing the traits of a neuroorganizational approach—but by default. There is one major and important difference between our examples and the way the brain itself is organized—the brain achieves excellence as a result of holistic activity and organization: it consists of a unified and interdependent whole that is greater than the sum of its constituent parts. It is our contention that the core attribute—almost the defining attribute—of a neuroorganizational approach is predicated on a holistic explication whereby an organization of any size and purpose deliberately chooses to deploy neuroanatomical principles as the basis for organizing its operations. It is not a "pick and mix" process whereby an organization selects only the bits it thinks are important to improve performance. Rather, it is more likely to be an "all or nothing" assignment, where a truly binary choice in favour of neuroorganizational principles could benefit organizational performance and ultimately survival.

Our goal in writing this paper has not been to provide a set of tools or devise a model ready for immediate implementation in organizations, nor to test such a model. That remains a long-term aim that will require considerable resourcing in future research before becoming a reality. Rather, what we have attempted to do is begin to identify areas of organizational theory and practice where the application of neuroanatomical principles and approaches may be beneficial to the development of high-performing organizations. The ideas and research questions we have identified in each of the five sections of our paper, which are by no means comprehensive or exclusive, are likely to serve as a basis from which to begin exploring the potential of this approach. Taken together, these constitute a core research agenda that focus attention on undertaking novel science-based and empirically-driven research into:

- "organizational mapping" based on neuroanatomical principles;
- the mutual interdependence of "doing" and "supporting" primary roles;
- the separation of high order "conscious" and routine "subconscious" functions;
- resourcing "developmental remodelling"; and
- organizational memory, plasticity and "creative intelligence."

Whilst it is likely that some of these will turn out to be more significant than others for organizational research and practice, it is our belief that they will lead to significant developments from both an academic and practical standpoint. Nowhere is this likely to be more applicable than in the area of strategic management, with its focus on strategic awareness, strategic analysis and the management of strategic change—what Stacey (2007) aptly refers to as managing the challenges of dynamic complexity. If strategic management is to face up to these challenges, as well as choosing and implementing strategy, then it does not take much of a leap of faith to recognise the potential contribution that each of these neuroanatomically informed elucidations could make to enhancing an individual organization's strategic awareness, improving its strategic analysis abilities and making its management of strategic change more efficacious. In addition, our neuroorganizational approach suggests that the conventional ways of exploring the linkages between strategy, strategic management and organizational structure may require a radical re-examination. For example, the discrete structural designs typically identified in the mainstream literature (c.f. Thompson & Martin, 2005) include variants of entrepreneurial, functional, divisional, holding company and/or matrix forms. Whilst it is possible to identify one or more elements of each in many of the high-performing organizations discussed in this paper—including Virgin Atlantic, Southwest, JetBlue and easyJet—it is also clear that, at the formal level of organizational design, there appears to be a missing element. What characterises each of these organizations is a focus on a somewhat different organizational structure: one that heightens co-operation over competition and empowerment over bureaucratic or functional control (c.f. Peters & Waterman (1982) –"simple form, lean staff," "productivity through people," "simultaneous loose-tight properties" and an obsession with delayering).

Although the focus of this paper has been on exploring the requirements for a neuroanatomical approach to organizational performance at the level of the individual organization, it is quite possible that such an approach could be transferred to other areas of social research in addition to those outlined previously (e.g., artificial neural networks, neuroeconomics, neuromarketing and organizational cognitive neuroscience). For example, the emergence of highly influential research under the aegis of the "new economic geography," "clusters" and the "new economics of competition"—where critical masses of unusual competitive success in particular sectors of economic activity are concentrated uniquely in local things like knowledge, relationships and motivation (e.g., Silicon Valley) (Porter, 1998; Porter, 2003)—are highly suggestive of the potential applicability of neuroanatomical analysis and exploration. For as Krugman (1998) has aptly put it, critics of economics have argued that the field takes too little account of a set of interrelated possibilities, such as the existence of cumulative processes of change involving "circular causation," the persistent effects of historical accident via "path dependence" and the occasional emergence of discontinuous change (maybe even "punctuated equilibrium"). For the most part, however, such efforts have been forced and relatively unconvincing. In the "new economic geography," however, such non-linear phenomena emerge absolutely naturally from the most basic models. One potentially exciting development in neuroanatomy that may prove to be particularly useful with regard to such modelling is the fledgling science of 'connectomics,' which aims to generate detailed maps of synaptic connectivity—network wiring diagrams—in the brain (Lichtman & Sanes, 2008). Such large-scale circuit reconstruction in the nervous system is getting ever closer, bringing with it numerous, transferable advances in computing technologies and expertise required to map, model and visualise extensive sets of connections and relationships.

In conclusion, a programme of truly interdisciplinary and collaborative research into neuroanatomical applications in organizational research—in areas as diverse as strategic management and organizational design (and possibly beyond)—may ultimately lead to business-ready tools that can be utilised by managers in aspirant high-performing organizations. To the best of our knowledge, this paper represents the first step in undertaking such interdisciplinary and collaborative research.

Acknowledgements

The authors would like to thank Drs Gordon Findlater, Fanney Kristmundsdottir and Simon Parson for assistance with preparing the figures, and Simon Bailey and Professor Violina Rindova for constructive comments on the manuscript.

References

Adams, D.A., Todd, P.A. and Nelson, R.R. (1993) A comparative-evaluation of the impact of electronic and voice mail on organizational communication. Information & Management 24:9–21.

Ambrose, D. (1995) Creatively intelligent post-industrial organizations and intellectually impaired bureaucracies. Journal of Creative Behavior 29(1):1–15.

Asaro, P. (2006) Working models and the synthetic model: Electronic brains as mediators. Science Studies 19:12–34.

Beer, S. (1972) Brain of the Firm: The Managerial Cybernetics of Organization. London: Allen Lane The Penguin Press.

Boillee, S., Yamanaka, K., Lobsiger, C., Copeland, N., Jenkins, N., Kassiotis, G., Kollias, G. and Cleveland, D. (2006) Onset and progression in inherited ALS determined by motor neurons and microglia. Science 312:1389–1392.

Brown, P. (2004) The evolving role of strategic management development. Journal of Management Development 24:209–222.

Bryman, A. (1992) Charisma and Leadership in Organizations. London: Sage.

Bryman, A. (2004) The Disneyization of Society. London: Sage.

Bryman, A., Gillingwater, D. and McGuinness, I. (1996) Industry culture and strategic response: The case of the British bus industry. Studies in Cultures, Organizations and Societies 2:191–208.

Buenstorf, G. and Witt, U. (2006) How problems of organisational growth in firms affect industry entry and exit. Revue de l'OFCE 97:47–62.

Butler, M.J.R. & Senior, C. (2007) Towards an organizational cognitive neuroscience. In: The Social Cognitive Neuroscience of Organizations (Eds: Senior, C. & Butler, M.J.). New York: Annals of the New York Academy of Sciences.

Camerer, C.F., Loewenstein, G. and Prelec, D. (2004) Neuroeconomics: Why economics needs brains. Scandinavian Journal of Economics 106(3):555–579.

Capodagli, B. and Jackson, L. (1999) The Disney Way. New York: McGraw-Hill.

Catani, M. (2006) Diffusion tensor magnetic resonance imaging tractography in cognitive disorders. Current Opinion in Neurology 19:599–606.

Christensen, C., Suárez, F. and Utterback, J. (1998) Strategies for survival in fast-changing industries. Management Science 44:S207–220.

Christopher, W. (2007) Holistic Management: Managing What Matters for Company Success. New York: John Wiley.

Ciborra, C. (2001a) A critical review of the literature on the management of corporate information infrastructure, (In) Ciborra, C. (Ed) From Control to Drift: The Dynamics of Corporate Information Infrastructures. Oxford: Oxford University Press.

Ciborra, C. (2001b) From Alignment to Loose Coupling: From MedNet to www.roche.com. In: Ciborra, C. (Ed) From Control to Drift: The Dynamics of Corporate Information Infrastructures. Oxford: Oxford University Press.

Combe, I. and Botschen, G. (2004) Strategy paradigms for the management of quality: Dealing with complexity. European Journal of Marketing 38:500–523.

Cooke, S.F. and Bliss, T.V. (2006) Plasticity in the human central nervous system. Brain 129:1659–1673.

Crossman, A.R. and Neary, D. (2005) Neuroanatomy: An Illustrated Colour Text (3rd Ed.). Oxford: Churchill Livingstone.

Dahlgaard-Park, S. and Dahlgaard, J. (2007) Excellence—25 years evolution, Journal of Management History 13:371–393.

Denoyelle, P. and Larreche, J-C. (1995) Virgin Atlantic Airways: 10 years later (and beyond). Case #595–023-1, INSEAD France, Fontainebleau.

Dewar, D., Yam, P. and McCulloch, J. (1999) Drug development for stroke: Importance of protecting cerebral white matter. European Journal of Pharmacology 375:41–50.

Disney Institute (2001) Be Our Guest: Perfecting the Art of Customer Service. New York: Disney Editions.

Edvardsson, B. and Enquist, B. (2002) "The IKEA saga": How service culture drives service strategy. Services Industry Journal 22:153–186.

Eppler, M. and Mengis, J. (2004) The concept of information overload: A review of literature from organization science, accounting, marketing, MIS, and related disciplines. The Information Society 20:325–344.

Feldman, R.M. and Feldman, S.P. (2006) What links the chain: An essay on organizational remembering as practice. Organization 13:861–887.

Ferguson, B., Matyszak, M.K., Esiri, M.M. and Perry, V.H. (1997) Axonal damage in acute multiple sclerosis lesions. Brain 120:393–399.

Fiorini, F. (2002) JetBlue pursues growth while staying "small." Aviation Week & Space Technology 156:41–43.

Flouris, T. and Oswald, S. (2006) Designing and Executing Strategy in Aviation Management. Aldershot: Ashgate.

Freedman, M. (2003) The genius is in the implementation. Journal of Business Strategy March/April:26–31.

Freeman, M.R. (2006) Sculpting the nervous system: Glial control of neuronal development. Current Opinion in Neurobiology 16:119–125.

Garud, R. and Kotha, S. (1994) Using the brain as a metaphor to model flexible production systems. Academy of Management Review 19:671–698.

Gittell, J. (2003) The Southwest Airlines Way: Using the Power of Relationships to Achieve High Performance. New York: McGraw-Hill.

Greiner, L. (1998) Evolution and revolution as organizations grow. Harvard Business Review 76:55–67.

Grinyer, P., Mayes, D. and McKiernan, P. (1989) Sharpbenders: The Secrets of Unleashing Corporate Potential. Oxford: Blackwell.

Hall, W. (1980) Survival strategies in a hostile environment. Harvard Business Review 58:75–85.

Harari, O. (1994) The brain-based organization. Management Review 83:57–60.

Hart, S. (1992) An integrative framework for strategy-making processes. Academy of Management Review 17:327–351.

Hart, S. and Banbury, C. (1994) How strategy-making processes can make a difference. Strategic Management Journal 15:251–269.

Hatch, M. and Cunliffe, A. (2006) Organization Theory: Modern, Symbolic, and Postmodern Perspectives (2nd Ed.). Oxford: Oxford University Press.

Hensch, T.K. (2005) Critical period plasticity in local neocortical circuits. Nature Reviews Neuroscience 6:877–888.

Johnson, G. (1988) Rethinking incrementalism. Strategic Management Journal 9:75–91.

Kaplan, R. and Norton, D. (2000) Having trouble with your strategy? Then map it. Harvard Business Review September-October: 3–11.

Kim, T., Oh, H. and Swaminathan, A. (2006) Framing interorganizational network change: A network inertia perspective. Academy of Management Review 31:704–720.

Kim, W.C. and Mauborgne, R. (1999) Strategy, value innovation, and the knowledge economy. Sloan Management Review Spring:41–54.

Klein, J. (1990) Interdisciplinarity: History, Theory and Practice. Detroit: Wayne State University Press.

Kotler, P. (1999) Kotler on Marketing. New York: Free Press.

Kriegstein, A., Noctor, S. and Martinez-Cerdeno, V. (2006) Patterns of neural stem and progenitor cell division may underlie evolutionary cortical expansion. Nature Reviews Neuroscience 7:883–890.

Krugman, P. (1998) What's new about the new economic geography? Oxford Review of Economic Policy 14:7–17.

Kumar, N. (2004) Marketing as Strategy: Understanding the CEO's Agenda for Driving Growth and Innovation. Boston: Harvard Business School Press.

Kuemmerle, W. (1999) Commentary 3: easyJet: Big ticket entrepreneurship. European Management Journal 17:36–37.

Larsen, M. and Leinsdorff, T. (1998) Organizational learning as a test-bed for business process reengineering, Proceedings of the 31st Hawaii International Conference on System Sciences, 6-9 January, 5:343–354.

Lee, N., Broderick, A. and Chamberlain, L. (2007) What is "neuromarketing"? A discussion and agenda for future research. International Journal of Psychophysiology 63:199–204.

Lester, D. (2004) Organisational life cycle stage and strategy: must they match? International Journal of Management and Decision Making 5:135–143.

Lichtman, J.W., Livet, J. and Sanes, J.R. (2008) A technicolour approach to the connectome. Nature Reviews Neuroscience 9:417–422.

Lichtman, J.W. and Sanes, J.R. (2008) Ome sweet ome: What can the genome tell us about the connectome? Current Opinion in Neurobiology 18:346–353.

Lievens, A. and Moenaert, R.K. (2000) Project team communication in financial service innovation. Journal of Management Studies 37:733–766.

Liker, J. (2004) The Toyota Way. New York: McGraw-Hill.

Lippitt, G.L. and Schmidt, W.H. (1967) Crises in a developing organization. Harvard Business Review 45:102–112.

Low, L.K. and Cheng, H.J. (2006) Axon pruning: an essential step underlying the developmental plasticity of neuronal connections. Philosophical Transactions of the Royal Society Part B 361:1531–1544.

Mankins, M. and Steele, R. (2005) Turning great strategy into great performance. Harvard Business Review 83:65–72.

Marshall, G.W., Michaels, C.E. and Mulki, J.P. (2007) Workplace isolation: Exploring the construct and its measurement. Psychology & Marketing 24:195–223.

Maurer, D. and Lewis, T.L. (2001) Visual acuity: The role of visual input in inducing postnatal change. Clinical Neuroscience Research 1:239–247.

Maurer, D., Lewis, T.L. and Mondloch, C.J. (2005) Missing sights: Consequences for visual cognitive development. Trends in Cognitive Sciences 9:144–151.

McCabe, K. (2003) Neuroeconomics. (In) Nadel, L., (Ed) Encyclopedia of Cognitive Science. New York: Macmillan.

McKnight, L., Vaaler, P. and Katz, R. (Eds) (2001) Creative Destruction: Business Survival Strategies in the Global Internet Economy. Boston: MIT Press.

Melis, M., Spiga, S. and Diana, M. (2005) The dopamine hypothesis of drug addiction: Hypodopaminergic state. International Review of Neurobiology 63:101–154.

Miller, D. and Friesen, P.H. (1980) Momentum and revolution in organizational adaptation. Academy of Management Journal 23:591–614.

Mitchell, A. (1999) Out of the shadows. Journal of Marketing Management 15:25–42.

Monteiro, E. and Hepso, V. (2001) Infrastructure strategy formation: Seize the day at Statoil. In: Ciborra, C. (Ed) From Control to Drift: The Dynamics of Corporate Information Infrastructures. Oxford: Oxford University Press.

Morgan, G. (2006) Images of Organization (Updated ed.). London: Sage.

Nagai, M., Re, D.B., Nagata, T., Chalazonitis, A., Jessell, T.M., Wichterle, H. and Przedborski, S. (2007) Astrocytes expressing ALS-linked mutated SOD1 release factors selectively toxic to motor neurons. Nature Neuroscience 10:615–622.

Nicholson, N., Rees, A. and Brooks-Rooney, A. (1990) Strategy, innovation and performance. Journal of Management Studies 27:511–534.

Nonaka, I. & Takuchi, H. (1995) The Knowledge-Creating Company: How Japanese Companies Create the Dynamics of Innovation. New York: Oxford University Press.

Pauwels, P. and Matthyssens, P. (2003) The dynamics of international market withdrawal. In: Jain, S. (Ed) Handbook of Research in International Marketing. Cheltenham: Edward Elgar.

Peters, T. and Waterman, R. (1982) In Search of Excellence: Lessons from America's Best-Run Companies. London: HarperCollins.

Peterson, B. and Glab, J. (1995) Rapid Descent: Deregulation and the Shakeout in the Airlines. New York: Simon and Schuster.

Porter, M.E. (1998) Clusters and the new economics of competition. Harvard Business Review Nov-Dec: 77–90.

Porter, M.E. (2003) The economic performance of regions. Regional Studies 37:549–578.

Quinn, R.E. and Cameron, K. (1983) Organizational life cycles and shifting criteria of effectiveness: Some preliminary evidence. Management Science 29:33–51.

Rae, D. (2001) EasyJet: A Case of Entrepreneurial Management? Strategic Change 10:325–336.

Schmahmann, J. D. (2003) Vascular syndromes of the thalamus. Stroke 34:2264–2278.

Senior, C. & Butler, M.J.R. (Eds) (2007) The Social Cognitive Neuroscience of Organizations. New York: Annals of the New York Academy of Sciences (1118).

Sherman, D.L. and Brophy, P.J. (2005) Mechanisms of axon ensheathment and myelin growth. Nature Reviews Neuroscience 6:683–690.

Smolensky, M. & Stein, E. (Eds) (1998) Human Factors in Air Traffic Control. San Diego: Academic Press.

Stacey, R.D. (2007) Strategic Management and Organisational Dynamics: The Challenge of Complexity to Ways of Thinking About Organisations. Harlow: Financial Times/Prentice-Hall.

Stadler, C. (2007) The four principles of enduring success. Harvard Business Review 85:62–72.

Straus, B. (2009) Neeleman out as JetBlue CEO. Air Transport World, 11th May. (Accessed March 2009: www.atwonline.com/news/story.html?storyID=8864)

Sull, D. (1999) Case study: easyJet's $500 million gamble. European Management Journal 17:20–32.

SWA (2009) Customer service commitment. Southwest Airlines, Dallas TX.

(Accessed March 2009: www.southwest.com/about_swa/customer_service_commitment/ customer_service_commitment.pdf)

Tegarden, D.P. and Sheetz, S.D. (2003) Group cognitive mapping: a methodology and system for capturing and evaluating managerial and organizational cognition. Omega 31:113–125.

Thompson, J.L. and Martin, F. (2005) Strategic Management: Awareness, Analysis and Change. (5th ed). London: Thomson Learning.

Trachtenberg, J.T., Chen, B.E., Knott, G.W., Feng, G., Sanes, J.R., Welker, E. and Svoboda, K. (2002) Long-term in vivo imaging of experience-dependent synaptic plasticity in adult cortex. Nature 420:788–794.

VAA (2009) Welcome to the Virgin Atlantic careers site. Virgin Atlantic Airways, Crawley. (Accessed March 2009: http://www.virgin-atlantic.com/en/gb/careers/index.jsp)

Van de Ven, A.H. (1980) Early Planning, Implementation, and Performance of New Organizations. In: The Organizational Life Cycle (Eds: Kimberly, J. & Miles, R.). San Francisco: Jossey-Bass.

Venkatraman, N. and Ramanujam, V. (1986) Measurement of business performance in strategy research: A comparison of approaches. Academy of Management Review 11:801–814.

Virgin Group (2009) About Virgin: What we are about. Virgin Group, London. (Accessed March 2009: http://www.virgin.com/AboutVirgin/WhatWeAreAbout/)

Vrecko, S. (2008) Capital ventures into biology: Biosocial dynamics in the industry and science of gambling, Economy and Society 37:50–67.

Wakana, S., Jiang, H., Nagae-Poetscher, L.M., van Zijl, P.C. and Mori, S. (2004) Fiber tract-based atlas of human white matter anatomy. Radiology 230:77–87.

Walsh, J.P. and Ungson, G.R. (1991) Organizational memory. Academy of Management Review 16:57–91.

Wilson, H., Daniel, E. and McDonald, M. (2002) Factors for success in customer relationship management (CRM) systems. Journal of Marketing Management 18:193–219.

Wishart, T.M., Parson, S.H. and Gillingwater, T.H. (2006) Synaptic vulnerability in neurodegenerative disease. Journal of Neuropathology and Experimental Neurology 65:733–739.

Wolpaw, J.R. (2006) The education and re-education of the spinal cord. Progress in Brain Research 157:261–280.

Womack, J., Jones, D. and Roos, D. (2007) The Machine That Changed the World: The Story of Lean Production—Toyota's Secret Weapon in the Global Car Wars That is Now Revolutionizing World Industry (2nd ed.). New York: Simon & Schuster.

Wyatt, R.M. and Balice-Gordon, R.J. (2003) Activity-dependent elimination of neuromuscular synapses. Journal of Neurocytology 32:777–794. Wynbrandt, J. (2004) Flying High. New York: John Wiley.

Neural Networks and their Application to Finance

Martin P. Wallace

ABSTRACT

Neural networks are one such process, that is, it maps some type of input stream of information to an output stream of data. It consists of ways to connect data/information to produce output that is consistent with the processes. It may seem simple, but as the analysis will highlight, this process is far from trivial. Today neural networks have been integrated into most fields and are a very important analytical tool. Neural networks are trained without the restriction of a model to derive parameters and discover relationships, driven and shaped solely by the nature of the data. This has profound implications and applicability to the finance field. These areas will be analyzed with specific examples in each area.

What are Neural Networks?

The human brain is a very complex part of the human body, due mainly to the interactions and connectivity with other parts of our body, and the way it controls and defines every aspect of our being. The brain has continued to be a mystery to many scientists, but its role and capacity to process information is mimicked in many aspects of academia. Neural networks are one such process, that is, it maps some type of input stream of information to an output stream of data. It consists of ways to connect data/information to produce output that is consistent with the processes. It may seem simple, but as the analysis will highlight, this process is far from trivial.

A neural network works in a similar methodological way to connect processing elements to produce results from a complex analytical study or principle that depends on many interconnected explanatory variables. According to Smith initially neural networks were characterized as a computer science phenomenon with uses (Smith para 2):

- processing elements
- a high degree of interconnectivity
- dependence of variables

The basic idea behind a neural network is presented in Figure 1 below; where there are different inputs that combined create an output, however the ratio is not one to one, since there may be interactions between inputs and more so backward linkages between output and input, as presented in the diagram below. The figure was adapted form Stergiou and Siganos (para 2) to highlight the similarity between processes in the brain and neural networking.

The History of Neural Networks

Neural networks were originally devised to understand the workings of the human brain (a formidable task). However, there developed a multidisciplinary trend with the constant interaction of researchers across disciplines who tried to apply the neurological activities of the brain with classifying computer programs and functions (Stergiou and Siganos para 5).

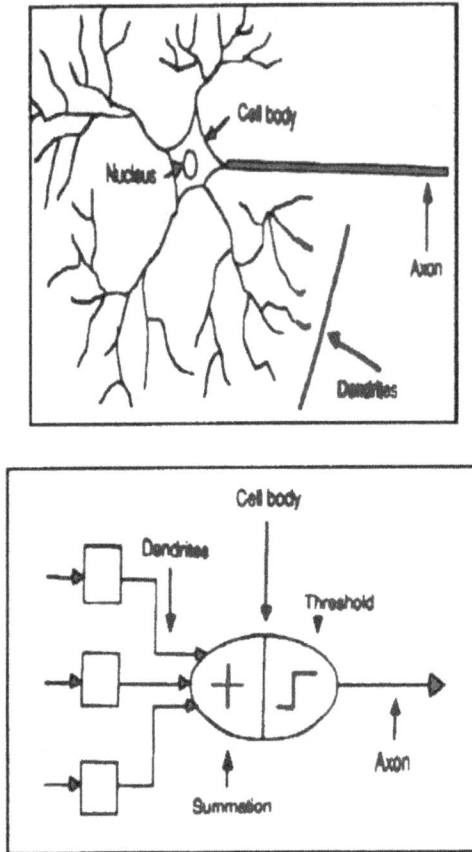

Figure 1: Neural Networking and Similarities with the Workings of the Human Brain

The first use and concept of neural networks began with linear classifications of the input-output relation represented generally equation 1 below:

$$Y = a + bx, \text{ with } x \in R^n, a \in R^n, b \in R^n \qquad (1)$$

Equation 1 was the typical development in neural networks and classified the general form of the Perceptron, which developed considerable interest and research in the 1950s (Stergiou and Siganos para 6). This model clearly has limitations, since it can only specify linear relationships in the input space; and will not classify complex data models that have a non-linear relationship.

Neural networks continued to advance and developed a multilayered algorithm that had the ability for bi-directional flow on inputs. Figure 2 below, highlights the general neural model that was adapted and transformed across disciplines. It highlights the development and historical progress of neural networks as its applicability across research arenas changed (Stasoft para 20).

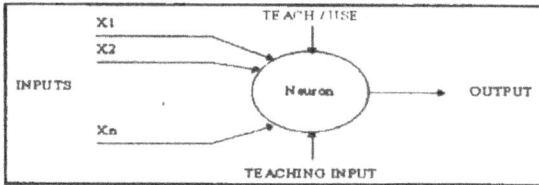

Figure 2: Multilayered Neural Network Model adapted from Stergiou and Siganos

Neural network development was not without its period of criticism and general disrepute. During this period, the flaws of the single layered Perceptron model were highlighted (like that shown in figure 1). This caused a general decline in funding and research associated with neural networks. Its overall use and computational ability to solve complex problems was questioned and this lead to limited use.

Nevertheless, neural networks have regained popularity and are being used in a wide array of fields within the natural and social sciences. Models such as those in figure 2 generally increased in complexity, the development of the Cognitron in 1975 with training and learning algorithm, along with the ability to change weights and interactivity across input sets in Rn, were developed and re-popularized the field. Other popular models such as the back-propagation network is which utilized a stochastic function to generalize the relationship and determine optimal parameters of a complex model via a more robust methodology (Stergiou and Siganos para 8). Equation 2 below highlights the complexity and development within the historical timeline that shows the sequential development of neural networks (note the difference between equation 1 and equation 2).

$$Y = h\,(a + bx),\ \text{with}\ x \in R^n,\ a \in R^n,\ b \in R^n \qquad (2)$$

where h is a logistic function.

Today neural networks have been integrated into most fields and are a very important analytical tool.

Why Use Neural Networks?

The recent increased interest and use of neural models stems primarily from its nonlinear models that can be trained to map past and future values of the input- output relationship. This adds analytical value, since it can extract relationships between governing the data that was not obvious using other analytical tools.

Neural networks are also used because of its capability to recognize pattern and the speed of its techniques to accurately solve complex processes in many applications. This is especially true of the backpropagation and Cognitron method introduced in the historical section of the paper. Neural networks help to characterize relationships via a nonlinear, non parametric inference technique, this is very rare and has many uses in a host of disciplines (Lendasse et al. 9).

Since a neural network is basically a data processing technique that links input streams with output, its use can be distinguished by four types of applications:

1. Classification of input stream

2. Association of output given sectors of input groupings

3. Codification of input by producing output within a reduced dimensional subspace

4. Simulation of output from input relationships and interconnections.

Neural networks offer the best 'back-drop' in which to extend simply methodologies to gain unique and extended results from models. The networks offer the added advantage of being able to establish a 'training' phase, where example inputs are presented and the networks learns to extract the relevant information from these patterns. With this, the network can generalize results and lead to logical and other unforeseen conclusions through the model.

Clearly neural networks surpass traditional models that use linear techniques and parameter threshold testing, hence neural networks add flexibility to the model. In addition, with the non-linear modeling capabilities, there are a wide range of complex models that can be easily implemented and analyzed.

Neural Networks Versus Conventional Computers

Neural networks have the unique capability of learning. That is, unlike conventional computers, sequences do not need to be dictated in order for the algorithm to be executed and produce meaningful results. This problem solving tools, creates a unique likeness to the human brain, that is, neural networks, use the interconnectedness of the elements of the model to arrive at logical and robust decisions, rather than follow a set of sequential steps, that may or may not solve the problem like computers do.

Neural networks also allow modeling and forecasting to be more efficient, why?

The necessary analytical framework provides an expansive model to analyze relationships that were not embedded in the methodology or mechanism used to solve the model. This highlights the a different aspect of model building, where the unique relationships between the variables creates the model, rather than trying to force variables to conform to a theoretical abstract that may or may not exist. Nevertheless, it is clear that neural networks cannot replace traditional computers, but can and will complement each other in problem solving mechanisms. Figure 2a below shows the generalized view of the multilayer perceptron network with specific emphasis on the multiconnectivity of the variables.

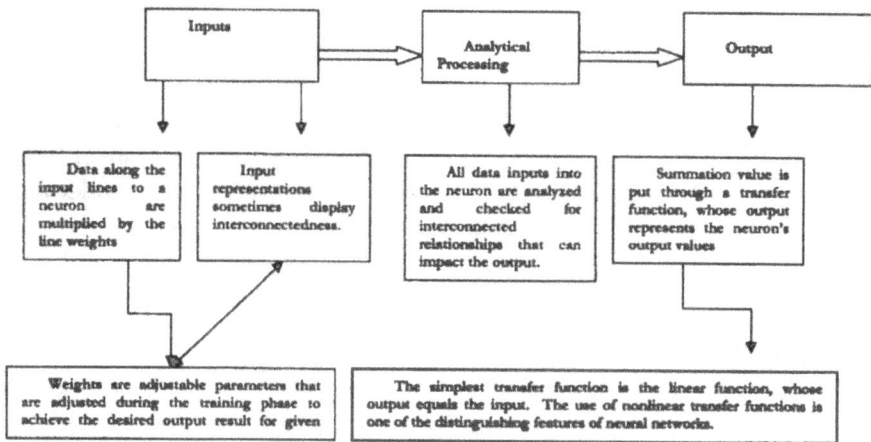

Figure 2a: Multilayer Perceptron Neural Network

Neural networks try to find the solution to problems by analyzing the variables and may come to unpredictable results, since the relationship between inputs and outputs is not specified within a particular methodology, but is rather loosely based on the unspecified steps to solve the problem. Conventional computers need this model or sequence of steps to solve problems, and as such will produce results that are stipulated by the model or framework used to analyze data.

Neural Networks in Finance

Neural networks are trained without the restriction of a model to derive parameters and discover relationships, driven and shaped solely by the nature of the data. This has profound implications and applicability to the finance field. These areas will be analyzed with specific examples in each area.

Time Series Analysis

Time series is a special form of data where past values may influence future values. Many financial models rely on understanding time series to adequately predict the functionality of financial markets and uses statistical inferences for forecasting purposes. The relationship between time variant variables in finance can be characterized by trends, cycles, and non-stationary behavior between data points that have serve a predictive or informational purpose to the model. Linear models have been used in the past to extract these relationships, but non-linear relationships exists between many financial variable, as such neural networks have a specific place within the financial literature and can be trained to map and future values of time series, so as to extract hidden structures and relationships that may govern the data (Lendasse et al. 5)

In discussing neural networks and time series analysis, it is beneficial to introduce the random walk Properties of pure random walk time series are of interest in providing a theoretical framework for financial time series and provides an applicable framework for neural networks in finance. Equation 3 below presents the random time series, which is used to model market prices.

$$P_t = P_{t-1} + u_t \tag{3}$$

Where p represents market prices, the ts subscripts are an index of time, and u is a stochastic variable, which is identically distributed. That is, $u \sim (0, c)$.

Typically, the random walk theory is applied to stock market analysis and is a useful background to the question of the nature of financial time series ("Financial Time Series as Random Walk" 6). Direct test of randomness within financial time series is fraught with problems and even the most advanced nonlinear models, still have not devised efficient ways to model the behavior of financial time series.

Based on the analysis above, the neural network seems like an appropriate model to analyze financial time series, since it will provide insight into the nature of the relationship between time series data (which can be useful for forecasting and stock market analysis which is examined below).

Figure 3 below shows, since the hypothesis being tested and debated in finance is whether financial time series have information that can be useful for predictive purposes, or just happen to follow a random walk. Neural networks have been useful in testing this hypothesis.

Figure 3: Stock Market Data from the New York Stock Exchange for Newmont Mining: Random Walk?

Stock Market Analysis

More individuals own stock more than ever. Stock pricing is now expansive and is an important aspect of financial economics. Therefore, many theorist look for different analytical tools to arrive at logical conclusions. Neural networks are technical models that can lead to insightful results and have a significant impact on the market.

A stock is generally considered over-valued if the price-earning ratio is high relative to the rate at which a company's earnings are likely to grow. The converse holds true for an under-valued stock. Because of the complexity and importance of valuing common stock, various techniques for accomplishing this task have been devised over time. The techniques that will be used encompass: 1) discounted cash flow valuation techniques, where the value of the stock is estimated based upon the present value of some measure of cash flow, including dividends, operating cash flow, and free cash flow; and 2) the relative valuation techniques, where the value of a stock is estimated based upon its current price relative to variables considered significant to valuation; 3) cost of capital; 4) capital budgeting.

The dividend discount model (DDM) is very useful for the stock market analysis and has been applied to the neural network in order to verify if entities are relatively stable and if prices are efficient and fair for stocks. DDM assumes that the value of a share of common stock is the present value of all future dividends.

$$\text{Value of stock} = \frac{\underline{D}_1}{(1+k)} + \frac{\underline{D}_2}{(1+k)^2} + \frac{\underline{D}_3}{(1+k)^3} + \frac{\underline{T}}{(1+k)^\infty} \qquad (4)$$

The inputs for the calculation include:

D_t = Dividends during period t

k = The required rate of return on stock j

T= terminal stock value

The analysis above is just a brief overview of the applicability of neural networks in the stock market. The random walk theory and DDM seemed like the most applicable (and popular) methodologies to analyze.

Capital Budgeting and Risk

Capital budgeting is one of the most important functions of financial management. It encompasses a process of planning expenditures on assets whose cash flows are expected to extend beyond one year. A company with growth rates and profit margins such as that are dictated by capital expenditure and investment cannot afford to ignore the importance of capital budgeting. Erroneous forecasts of asset requirements can have serious consequences, Therefore there is always a need for complex and accurate models to dictate the relationship between variables. How is capital budgeting associated with the neural networks? Capital budgeting typically involves a large amount of money, therefore when companies contemplate major capital expenditure programs, financing has to planned in advanced, hence the importance of stock value and forecasting mechanisms, as shown from the previous analysis, neural networks are important to this overall process. What is clear is that there is a direct link between capital budgeting and stock valuees. The more effective the firm's capital budgeting procedures, the higher its stock price. These arre hypotheses that are also tested via neural networks.

Once a potential capital budgeting project has been identified, its evaluation involves thhe same stteps that arre used in security analysis. Decision rules can be summarized by the fact that if the present value of the cash flows exceeds the cost the project is accepted. Otherwise, it should be rejected. (Alternatively, if the expected rate of retturn on the project exceeds itss cost of capital, the project is accepted). With this similarity, it is also relatively easy to use neural networks for forecasting and arriving at relationships and estimates between the variables.

Risk analysis is best approximated with market risk, that is, thhe part of aa project's risk that cannot be eliminated by diversification; it is measuured by the beta coefficient.

With the cost of equity equation analyzed in previous sections, it is not surprising that as market risk increases, the cost of equity increases aand stock value falls. Calculated risk coefficents and interactions withh other variables in finance can also be approximated via neural networks.

Generally, these applications rely on the fact that neural network models can be used to devise a function from observations (that may or may not have existed before). This is usually within the finance field where data is too complex to analyze (Smith para 6). Specifically the use of artificial intelligence techniques, which is generalized in figure 4 below is used within the financial industry with the latest methodological approaches aimed at maintaining the competitive edge.

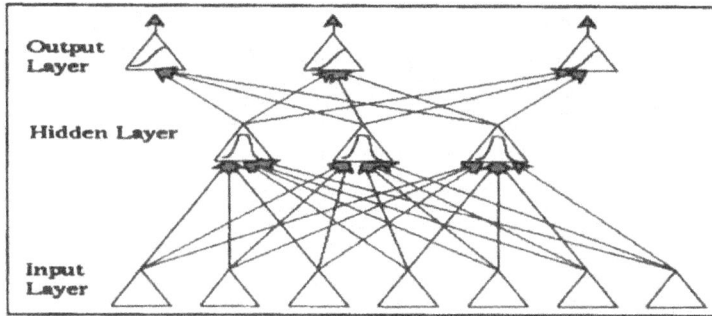

(Inclusive of Capital Budgeting and Risk Analysis)
Extracted from Leslie Smith, Centre for Cognitive and Computational Neuroscience

Figure 4: General Modeling of Neural Networks for Financial Capital Markets

Financial Forecasting

Neural networks provide forecasts of market prices and actions. These can then form the basis for trading the market in an automated system. A pre-trained network is the natural choice for real-time trading. The implementation of forecasts requires a strategy for dealing with adverse market moves; the question of when to enter or exit the market is also largely determined by forecasts, hence neural networks always have a role in finance.

There are a number of considerations in using neural networks for financial forecasting, however the neural network has an advanced pattern recognition technique, which makes it particularly useful in time series forecasting.

Neural networks have also been used to analyze rather profound hypotheses. The efficient market hypothesis states that if a market is considered efficient, than prices fully reflect all the relevant information, and buying and selling stock for capital gain is purely a matter of luck, rather than sound investment skills. Neural networks have been used to chart the relationship between financial forecasting, especially for the stock market and to test the relevance of the efficient market theory (Smith para 19).

The Future of Neural Networks: A Critical Review

It is argued that neural networks cannot do anything that cannot be done using traditional techniques, but have simplified the process of completing otherwise complex and arduous tasks that researchers and analyst once had to do. Therefore, there exist many areas that can use neural networks to both increase efficiency and accuracy or as a way to improve the general analysis of the models. They include and are not limited to investment analysis, to predict stock currencies beyond the simple linear market, as a mechanism for comparing signatures with those stored, for process control, engineering applications, and in marketing for advertising and promotions. It is clear that neural networks have a rather expansive application base and will provide useful analytical results to users (Stasoft para 44).

Nevertheless, neural networks are highly technical and require a great deal of expertise to implement, although computers exist to run programs and generate results, these models are highly complex and require a great deal of technical expertise, as such there use is limited within the 'real world,' although they have shown to be great theoretical models, outside of the natural science field (Smith para 34).

Some theorists have even argued that the use of neural networks undermines scientific knowledge, since it's applicable in other fields and highly technical nature may cause users to extract results without necessarily understanding the methodology used by the model.

Other areas of expansion within the neural network movement includes the production of a learning chip, sensory and sensing applications, new opportunities for forecasting stock and financial markets, as well as other financial and economic time series, the use in incomplete data to find relationships that exists, and extensions in neuroscience and biological neural networks (Stasoft para 55).

In conclusion recent developments in neural networks highlight the new opportunities that it provides as an analytical tool. The mathematical content of the methodology seems rather erudite and restrictive; however, this does not mean the neural networking is not a very analytical tool that can produce 'real' results, irrespective of the complexity of the methodology used.

Works Cited

"Financial Time Series as Random Walks." Extracted on March 5, 2007 from http://www.cs.sunysb.edu/~skiena/691/lect ures/lecture8.pdf

Gordon, Myron. The Investment, Financing, and Valuation of the Corporation. Irwin, 1962

Lendasse, A., Bodt, E., Wertz, V., and Verlesen, M. "Non-Linear Financial Time Series Forecasting—Application to the Bel 20 Stock Index." European Journal of Economic and Social Systems, 14 (1), 2000, 81-98. Retrieved on March 5, 2007 from http://www.edpsciences.org/articles/ejess/ pdf/2000/01/verleyse.pdf?access=ok

Smith, Leslie. "An Introduction to Neural Networks." Centre for Cognitive and Computational Neuroscience. April 2, 2003. Retrieved on March 6, 2007 from http://www.cs.stir.ac.uk/~lss/NNIntro/InvSlides

Stasoft Incorporated. "Neural Networks." Stasoft.Com, 1984-2003. Retrieved on March 6, 2007 from http://www.statsoft.com/textbook/stneunet .html#index

Stergiou, Christ's and Siganos, Dimitrios. "Neural Networks." Surprise, 4, 11, 1996. Retrieved on March 6, 2007 from http://www.doc.ic.ac.uk/~nd/surprise_96/journal/vol4/cs11/report.html

Implementing the Activity Base Costing System: A Case Study on Dakota Office Supply

Betty W. Steadman

OVERVIEW

Activity Based Costing (ABC) is an accounting method that allows an organization to determine actual costs associated with each product and/or service produced by the organization without regard to the organizational structure or other extraneous function. For Dakota Office Products (DOP), its existing costing system was inadequate because it is incapable of accounting for even all of the known costs such as the desktop delivery service as well as hidden costs such as the 10% DOP paid to maintain its working capital line of credit for accounts receivable (Kaplan, 2003, p.4). Since ABC is a powerful tool for measuring performance, identifying, describing, and assigning costs to, and reporting on an organization's operations it could solve much of

DOP's critical cost oversights (Caplan, Melumad & Ziv, 2005). Used holistically ABC can be utilized to also improve processes and identify opportunities to improve business effectiveness and efficiency by determining the true or real costs of a given product or service. ABC principles are used to focus management's attention on the total cost to produce a product or service, and as a basis for full cost recovery of a production or service process.

Situational Analysis

DOP is a regional office supply company with a strong reputation for customer service and quality supplies. Additionally, DOP is unafraid to adopt new service operations such as its "desk top" delivery option which delivered smaller orders directly to individual sites as we all as its traditional commercially delivered mass orders to customer distribution sites (Kaplan, 2003, pp.1-2). Additionally DOP deployed an Electronic Data Interchange (EDI) solution in order to ease data and payment transfer from and to customers as well as building a customer website that acts an order and account interface for its customers. Together these initiatives all expanded DOP's customer service and quality metrics but also came with a cost which the company had difficulty identifying. While exact costs were difficult to ascertain for DOP it is clear that the company is incurring expenses in a manner that it previously had not since its EBIT (earnings before interest & taxes) for fiscal year 2000 revealed a—1.3% loss (Kaplan, 2003, p.4). Since the company did not actually break down costs in order to arrive at a more accurate pricing schedule but instead relied only on a universal 15% markup over basic material costs, DOP had absolutely no way of being able to identify where the cost inefficiencies were in its operations.

Activity Based Costing

Traditional product costing such as DOP employs was designed for an industry comprised of companies having one-type (homogenous) of product, large direct (fixed) costs, very few data collection techniques and fewer data analysis tools, and low below line costs. By contrast, contemporary industry exhibits a highly diversified product/service mix as evidenced by DOP's employment of both commercial delivery methods and its new customized individually oriented delivery model accompanied by high overhead costs as compared to fixed labor costs, a mass amount of data with countless ways to analyze it, and considerable non-product, indirect, or hidden costs affecting overall pricing, costs and expenditures.

ABC is a system of accounting based on the allocation of resources and accounting for the value of those resources in relation to a specific task or activity. In this sense, labor and direct material costs lose the prominence that they are given in traditional costing systems and are accounted for as any other resource is accounted for. Activities such as sales, support, or other business processes are largely excluded. This is not the case in ABC because all the various functions, activities, and support processes of a company are broken down into pools of resources with a particular value attached to them. For example, developing a cost-driver rate for DOP would involve basing an estimate on an approximate 80% capacity rate to employees. For example, a cost-driver for DOP related to warehouse operations is 4,400,000 annually which would be divided by the number of cost-drivers such as order set-up, order entry, order validation, receiving, handling, packaging, loading, and delivery and the result would be the cost per occurrence of the cost-driver rate: $550k. This figure is then divided by the total number of instances these actions occurred which, as reported by DOP was 80k during 2000 and so the cost-driver the year would be $6.87 per activity on average. This figure would be used to identify unprofitable accounts such as the customer A and B mentioned in the case where customer A and B both generated approximately $100k in annual revenues and based on customer A's 21.2% markup this would mean that $78,800 represented the COGS (cost of goods sold) while the cost-driver would indicate that $10,992 went towards the associated activity costs which would mean that customer A's profitability is $10,208. By contrast, customer B's profitability is lower. Based on the same assumptions customer B's COGS is $77,600 while the cost-driver would be factored with an extra activity per carton and thus would be $12,366 plus the $3,000 to carry its accounts receivable balance resulting in a profitability of $7,034. Because of the additional activities associated with customer B and the hidden cost of carrying its accounts receivable balance, customer B is less profitable with the same overall revenues as customer A. While it is difficult to identify limitations in this ABC methodology with respect to these particular customers it could be argued that the goodwill created by the additional services for customer B will result in longer service and greater orders in the future. Recommended Application at DOP.

ABC in Practice at Dakota

Before performing ABC, a baseline or a starting point is needed for business process improvement and a baseline can be expressed in some form of model. This baseline is critical for DOP because in order to establish this baseline metric the analytics just performed must be done for each individual account. If DOP performs this activity on each customer the strategic management benefits would be substantial because all the excess cost-drivers could be eliminated resulting in much wider operating margins and thus profitability without increasing costs or committing resources to gain this efficiency. Therefore, a baseline is a documentation of the organization's policies, practices, methods, measures, costs and their interrelationships at a particular location at a particular point in time (Maiga & Jacobs, 2003). Through base-lining, activity inputs and outputs across functional lines of business can be identified. ABC is the only improvement methodology that provides output or unit costs. Value added activities are those for which the customers are usually willing to pay in some fashion for the product or service. Non-value added are activities that create waste, result in a delay of some sort, and potentially adds costs to the products or services. Resources are assigned to activities so that the activities can be performed in the first place. Some of Pilgrims' resources are measured in man-hours, machine hours as well as machine maintenance and operational overhead. It is through ABC that an organization can begin to see actual dollar costs against individual activities, and find opportunities to streamline or reduce those costs, or even eliminate the entire activity thus removing the cost altogether. This is the process inherent in ABC that reduces overall expenditures of the company.

Procedural Steps of ABC

Some typical steps in implementing an ABC program are often defined as five activities that must occur if a true cost accounting is to take place within the confines of an ABC operation. These five activities are typically listed as (Latshaw & Cortese-Danile, 2002): 1) Analysis of activities, 2) Cost gathering, 3) Associating costs with activities, 4) Base-lining output metrics, and 5) Cost analysis. These steps should be performed by an integrated team of cost accountants, floor managers, and project managers that have been committed by top management to work on the ABC project within DOP. Yet, the cost savings could be enormous. For example, by identifying the previous cost-drivers it is apparent that a simple migration of all customers over to the internet based ordering and billing system approximately 9,500 man hours can be removed from the current order fulfillment process resulting in a cost savings of at least $142,500 in labor alone not to mention efficiencies gained in the order fulfillment process where the cost-driver rate could be reduced by the removal of an activity which would amount to approximately $515,250 or a total cost savings of $657,750 annually. Considering DOP missed its earnings target by (470,000) this cost savings would actually return the company to profitability.

References

Caplan, D., Melumad, N. D., & Ziv, A. (2005). Activity-Based Costing and Cost Interdependencies among Products: The Denim Finishing Company. Issues in Accounting Education, 20(1), 51+.

Kaplan, R. (2003). Dakota Office Products. Harvard Business School, 9-102-021.

Latshaw, C. A., & Cortese-Danile, T. M. (2002). Activity-Based Costing: Usage and Pitfalls. Review of Business, 23(1), 30+.

Maiga, A. S., & Jacobs, F. A. (2003). Balanced Scorecard, Activity-Based Costing and Company Performance: An Empirical Analysis. Journal of Managerial Issues, 15(3), 283+.

Steadman, B.W.—Case Study 4: Implementing the Activity Base Costing System: A Case Study on Dakota Office Supply

Profitability of the Greek Football Clubs: Implications for Financial Decisions Making

Dimitropoulos E. Panagiotis

ABSTRACT

In the present study we examine the profitability of the football clubs participating in the first division of the Greek Football League, as well as the factors that contribute to this performance, over the period from 1994 to 2004. The results indicated that the profitability of the football clubs is positively associated to their short run success, but not on the long run success and seasonal uncertainty of the league. Additionally, the size of the club, measured as a fraction of the club's assets, is a distinct factor which affects the financial performance positively. Finally, the level of asset turnover and ROA reported by the clubs proved to have a significant positive impact on profitability suggesting that those football clubs that are able to use their assets efficiently, are more resourceful by means of profitability.

Introduction

Soccer has become a capital market, the main characteristic of which is investment of uncounted billions. Extravagant expenditure for transfers, astronomical sums for signing of contracts with footballers, disputes and battles among sponsors to get 'star' footballers to promote and advertise their products, endless negotiations to obtain a share of the TV rights, professional managers and finding the model team for potential investors piece together the current soccer environment.

There are some indications with regard to the relationship between soccer and its macroeconomic consequences for the global economy. Certainly, the consequences are not as wide-ranging as those of a petroleum crisis in leading the economy out of a recess into a boom or vice-versa. One certain thing is that the impacts of a great soccer event are felt on an economy, even to a small extent. ABN-AMRO investigates these impacts in its Soccernomics 2006 report on the occasion of the World Cup that took place in Germany last July. Data from the past show that the World Cup winning country enjoys, on average, an extra bonus of 0.7% in its economic growth rate. On the other hand, the country whose team loses in a World Cup Final also loses an average 0.3% from its annual economic growth rate. Now, with regard to companies, the income of the 20 largest European clubs reached record levels in the period 2004/05, for the first time exceeding the barrier of 3 billion euros (Source: Deloitte, Soccer Money League 2006). A 6% growth rate on an annual basis has been shown in the income of the football 'giants' for the said period, exceeding many growth rates of other significant economic sectors.

In Greece, with the exception of 2004, when the Greek national team won the European Cup, no significant achievements have been recorded on an individual club basis. This is probably the main reason why the overwhelming majority of the teams appear to be in difficult financial straits, having accumulated great losses. Our main aim in this paper is to investigate the distinct factors (athletic and financial) that are related to their financial performance of Greek teams. To our knowledge, this is the first study within the Greek football setting which tries to distinguish the significant determinants of profitability, thus our study adds to a growing body of research on sports finance. The remainder of the paper is organized as follows: In the next section we provide a short overview of the relevant regulatory framework. Section 3 develops the theoretical background while in section 4 there is a description of the data and the methodology employed. Section 5 presents the empirical results. Section 6 concludes with a summary.

Regulatory Framework

Professional soccer in Greece is governed by the rules and provisions enforced by the Ministry of Development. The competent body for the supervision and observance of the regulatory framework as well as the smooth running of the Super League is the Hellenic Football Federation. After 2003, and in response to the innumerable irregularities ascertained in this sector, a new inspection body was established in the form of the Professional Athletics Committee, which is now, as an independent authority, the main inspector and principal guarantor of transparency and legality in the field of professional athletics. The Athletic Football Clubs have financial and administrative independence. They keep minute-books of the decisions of the boards of directors and third class accounting books as per the Greek Code for Accounting Books and Records and draw up a budget and balance sheet. The Athletic Football Clubs are obliged to submit to the Professional Athletics Committee, at least 15 days before the season begins, an income and expense budget for the new season, balanced and verified by at least one certified public accountant. Nevertheless, significant problems with regard to the mismanagement of club finances have remained unsolved for decades. Thus, in accordance with recent data, the accumulated accounting losses have exceeded the share capital of the Athletic Football Clubs and their operation should normally have been curtailed. However, the paradox is that, with the tolerance of the authorities and under the legal shield provided by ARTICLE 44, LAW 1892/90, soccer teams continue to compete in the championships, while their debts are either prescribed or settled by long-term installments which run the risk of never being collected.

Certainly, there is no uniform Community legislation to determine what should be applied in every European football league. Each country has its own federation that determines its own rules and penalties in cases of debt. The strictest countries in respect of debts are France, Italy, Austria, Denmark, Hungary and Switzerland. More specifically, in Italy—and the same stands in Belgium—when the debt of an Athletic Football Club exceeds a specific amount, the team is relegated to a lower division. In Switzerland, when a soccer team has debts, it is deprived of its professional status and is relegated to the 1st amateur division. In Spain, a case of relegation due to debts has never been recorded. In England, a team that has debts loses ten points until they have been settled (which rarely happens more than twice in one team, since it is ultimately deprived of its professional status should it not comply). In any case, the current situation in Greece favours the phenomenon that the majority of teams, either to a large or small extent, show fictitious losses, which at all events are written off by the State, in order to avoid paying taxes.

Theoretical Background

Research on sport economics has been extended on a variety of issues and fields of sports during the last three decades. Since the seminal work by Sloane (1969, 1971) many researchers have attempted to examine the special features of the economics of the professional team sports. Bird (1982), Cairns et. al (1986) and Dobson and Goddard (1992, 2001) among others, have tried to determine the characteristics of the demand function for sport events and the factors that affect the market equilibrium. Under this framework, Janssens and Kesenne (1987), Jennett (1984), Peel and Thomas (1988) extended the aforementioned work and introduced match and seasonal uncertainty as distinct factors that affect the demand for sport events. The intuition behind the above mentioned argument is that the uncertainty of an outcome is positively associated to match attendance since spectators generally prefer a close match than a one-sided match (Borland and Lye, 1992).

Nevertheless, it was in the beginning of the 1990's that the researchers' interest shifted to the increased importance of financial management for the viability and prosperity of professional teams. Two distinctive studies by Szymanski and Smith (1995, 1997) examined the impact of market and wage size on the financial performance of the English football clubs. They documented that the English football industry is a mature industry with declining demand, including many loss reporting teams which assets are under-utilized and require increased investment funds in order to improve the quality of their product and to meet governmental safety standards.

Moreover, a relative study conducted by Burger and Walters (2003) addressed the issue of market size and the financial performance of the teams participating on the US major league baseball. Their findings document that market size and the team performance are strongly associated with marginal revenues and thus teams' motivation to bid for new talent players. Finally, a recent paper by Pinnuck and Potter (2006) carried out within the Australian Football League examined how the on-field football success impacts on the off-field financial performance of the AFL football clubs. Their results provide evidence that the short and long run success of the clubs and the uncertainty of the outcome affect attendance at AFL matches. Additionally, they argue that changes in membership are a positive function of past success and found to be directly associated to marketing expenditures made by the club.

However, one important limitation of all the aforementioned studies is that they have restricted their research agenda trying to determine the relation between sport characteristics (athletic success) and revenue creation. No research until now has properly addressed the issue of profitability and the specific factors that contribute to profit making. The aim of this paper is to enrich the existing literature by attempting to identify the effects of firm specific characteristics (size, financial risk, liquidity, cash flows) on the profitability of the football clubs.

Data & Methodology

Data Selection Procedure

The sample of our study comprises data from 17 football clubs participating on the first division of the Greek professional football league over the period from 1994 to 2004. We chose the specific period of investigation for two reasons. First of all, we needed to be sure that we had enough firm-year observations in order to conduct the time-series tests and secondly because we wanted to control for any bias that maybe exist on the financial data due to unexpected success of the Greek national football team on the Euro 2004 championship. Each football club had to meet some specific criteria in order to be included in the sample:

1. Each club must have full financial data of earnings, assets, liabilities and cash flows published in their annual financial statements, audited by an independent chartered accountant.

2. Each club must have sufficient data regarding their on-field performance and specifically, league position, wins achieved in a season, league scores and tickets sold.

3. Each club must have participated at least once in the 1st division of the professional football league.

All financial data were extracted from the I.C.A.P. database while data concerning the clubs athletic success were hand collected from the website of the Greek Football Federation (E.P.O.). No further trimming on data was conducted since we did not want to limit our sample to a small number of clubs because this could bias the final results.

Model Specification for Evaluating Profitability

In this section we attempt to identify the effect of firm specific characteristics on the profitability of the football clubs. However, there is no previous research on this type of business organization that has properly addressed the issue of profitability and the specific factors that contribute to profit making. Therefore it is necessary to examine additional financial and firm quality variables in the context of the following model:

$$PR_{it} = \gamma_0 + \gamma_1 SIZE_{it} + \gamma_2 LEV_{it} + \gamma_3 LIQ_{it} + \gamma_4 CF_{it} + \gamma_5 AT_{it} + \gamma_6 ROA_{it} + \gamma_7 ROE_{it} + \gamma_8 WIN_{it} + \gamma_9 PASPOS_{it} + \gamma_{10} UNCERT_{it} + v_{it} \quad (1)$$

PR is the ratio of earnings to sales, SIZE is the natural logarithm of total assets, LEV is the ratio of total debt to equity, LIQ is the ratio of current assets to current liabilities, CF is the ratio of cash flow to total assets, AT is the asset turnover ration measured by net sales over total assets, ROA is the returns on assets estimated as net income over assets and ROE is return on equity estimated by dividing net income over shareholders' equity. WIN is the number of wins achieved by the team in every season, PASPOS is a dummy variable taking the value of (1) if the team has finished in the six first positions of the ladder in the previous three periods and zero otherwise, UNCERT is the league's uncertainty measured by the ratio of the points that each team lags from the champion divided by the total points of the champion team.

As Majundar (1997), and Barbosa and Louri (2005) argue, firm size impacts significantly on firm level performance. Large firms may be able to generate superior performance since they can exploit economies of scale and organize their activities more efficiently resulting into increased profit streams compared to small firms. Thus if this intuition is valid we expect large clubs (according to total assets) to be more efficient by means of profitability and this will be depicted by a positive coefficient on the SIZE variable.

Additionally, firm specific choices are closely related to financial risk and the asset management efficiency may lead into heterogeneity within the industry which can help explain firm performance (Copeland and Weston, 1983). In order to control for financial risk that can be associated to the club's financial performance we introduced the variables of leverage, liquidity and the level of cash flows per assets. A positive and significant coefficient on either LIQ or CF indicates that football clubs are able to convert assets into cash thus resources can be used quickly so as to respond to profit opportunities. Furthermore, leverage is an indicator of the risks associated with the probability of default by the firm and as Penman (2001) argues the lower the leverage ratio the greater the financial security and the higher the level of the expected profits. Thus according to the previous discussion we expect to find a negative and significant coefficient on the LEV variable.

Moreover, the asset turnover and return on assets (ROA) variables are introduced in order to control for the level of efficiency in assets management. It is expected that the higher the ratio of sales over assets and ROA the higher the profitability, since clubs are more able to differentiate their product resulting into increased profits. Therefore, a positive and significant coefficients is expected on the AT and ROA variables. The last financial variable employed in our study is return on equity (ROE) or how well managers are employing the funds invested by the football club's shareholders in order to generate returns. ROE is the most comprehensive indicator of a firm's performance thus we expect a positive and significant coefficient on the ROE variable.

Also, prior research by Forrest et. al (2002), Forrest and Simmons (2002), Borland and Macdonald (2003), document that profitability is significantly affected by the short and long run success of the football club and the uncertainty of the outcome. Under this framework the variable we use in order to control for the recent athletic success of the football clubs is the number of wins (WIN) the team has achieved during the season. If there is a positive association between attendance and the short run success of the team then we expect to find a negative and positive coefficient on WIN variable respectively.

Additionally, profitability may also be affected by the long run success of the football club since we can assume that the athletic performance persists over seasons thus a team which is successful on the long run is more possible to be also successful on the short run which in turn leads into an increase in attendance (Pinnuck and Potter, 2006). The long run success (PASPOS) of the football clubs is controlled by using a dummy variable which takes the value of one (1) if the team has finished within the six first positions of the ladder and zero (0) otherwise. The reason for the aforementioned definition is that in Greece the first six places of the 1st division league give the ability to the runner ups to participate into the major European football events, Champions League and UEFA cup and this could result into increased revenue either by the UEFA organization, or by sponsors, TV licensees and tickets. Consequently if there is a relation between the long run success of the clubs and attendance this will be depicted by a significant and positive coefficient on the PASPOS variable.

Finally, Jennett (1984) argues that the uncertainty of the match outcome impacts positively on profitability since the closeness of the competition attracts more spectators. In order to control for the seasonal uncertainty (UNCERT) we constructed a ratio of the points that each team lags from the champion at the end of the season, divided by the total points of the champion team. The smaller the aforementioned ratio is, the higher the uncertainty of the championship which in turn will result into increased attendance. Thus if this intuition is valid we will expect a negative and significant coefficient on the UNCERT variable.

Empirical Results

Descriptive Statistics & Correlations

The following Table 1 includes the descriptive statistics of the sample variables for the whole period of investigation from 1994 to 2004. Regarding the on-field variables Greek football clubs achieve 13 wins during a season and the overall championship can be characterized by moderate uncertainty (0.43). As for the accounting variables we can argue that football clubs in the Greek professional league suffered from severe losses throughout the period of investigation, since the median net profit margin is negative and up to—1.10. Also the Greek football clubs are small in size, highly leveraged (1.29) and face intense liquidity problems since their current assets cover only the 26 per cent of current liabilities. Finally, the median AT value of 0.31 indicates that professional football clubs do not use their assets productively in order to generate sales and this fact is depicted in the median values of ROA and ROE which are negative and up to 16 per cent and 18 per cent respectively. Put it another way, from each euro of club's assets and equity, managers yield 18 cents net loss a fact that indicates severe financial mis-management.

Table 1: Descriptive statistics of the sample variables (1994-2004)

Variablies	Mean	Median	St. Deviation	1st Quartile	3rd Quartile
WIN	14.01	13.0	6.59	9.0	19.0
PASPOS	0.36	0.00	0.48	0.0	1.0
UNCERT	0.36	0.43	0.24	0.14	0.56
PR	2.93	-1.10	4.46	-3.44	0.55
SIZE	6.10	6.16	0.62	5.59	6.61
LEV	2.59	1.29	3.27	0.88	2.79
LIQ	0.44	0.26	0.68	0.16	0.56
CF	1.72	0.91	3.01	0.37	2.39
AT	0.45	0.31	0.42	0.16	0.57
ROA	0.28	-0.16	0.52	-0.47	0.014
ROE	-0.52	-0.18	1.81	-0.44	0.019

Sample comprises from 17 football clubs participating on the 1st division of the Greek football league from 1994-2004. WIN is the number of wins achieved by the team in every season, PASPOS is a dummy variable taking the value of (1) if the team has finished in the six first positions of the ladder in the previous three periods and zero otherwise, UNCERT is the league's uncertainty measured by the ratio of the points that each team lags from the champion divided by the total points of the champion team, PR is the ratio of earnings to sales, SIZE is the natural logarithm of total assets, LEV is the ratio of total debt to equity, LIQ is the ratio of current assets to current liabilities, CF is the ratio of cash flow to total assets, AT is the asset turnover ration measured by net sales over total assets, ROA is the returns on assets estimated as net income over assets and ROE is return on equity estimated by dividing net income over shareholders' equity.

Table 2 presents the correlation coefficients among the sample variables for the period under investigation. Pearson coefficients are below and Spearman co-efficients are above the diagonal. Regarding the on-filed performance variables, there are positive correlations between profitability and the WIN and PASPOS variables (0.41 and 0.31 in Pearson coefficients and 0.33 and 0.29 in Spearman coefficients respectively) suggesting that the short and long run of a football club enhances its profitability. Also the league's uncertainty impacts positively on the club's profitability since the correlation coefficient between PR and UNCERT is negative and significant on both types of estimates (Pearson coefficient is—0.37 and Spearman coefficient is—0.32 both significant at the 0.01 level). As for the financial variables PR is positively correlated with SIZE and AT (0.34 and 0.41 respectively on Pearson estimates and 0.23 and 0.70 on Spearman estimates, all significant at the 0.01 level) suggesting that larger clubs by means of total assets and clubs with greater asset utilization can achieve increased levels of profitability. Also profitability found to be positively and significantly correlated with liquidity (0.22 significant at the 0.01 level) and negatively correlated with leverage (-0.19 significant at the 0.05 level).

Table 2: Correlation coefficients among the sample variables (1994-2004)

Variables	WIN	PASPOS	UNCERT	PR	SIZE	LEV	LIQ	CF	AT	ROA	ROE
		0.74	-0.89	0.33	0.59	-0.018	-0.17	-0.29	0.10	-0.21	-0.28
WIN	1.0	(0.00)	(0.00)	(0.00)	(0.00)	(0.84)	(0.060)	(0.00)	(0.26)	(0.016)	(0.00)
PASPOS		1.0									
	(0.00)										
UNCERT	-0.87	-0.75	1.0	-0.32	-0.56	0.079	0.13	0.24	-0.09	0.25	0.26
	(0.00)	(0.00)		(0.00)	(0.00)	(0.38)	(0.13)	(0.00)	(0.32)	(0.006)	(0.00)

Table 2: *(Continued)*

Variables	WIN	PASPOS	UNCERT	PR	SIZE	LEV	LIQ	CF	AT	ROA	ROE
	0.41**	0.31**	-0.32**		0.22*	-0.14	-0.07	-0.17	0.70**	0.10	0.11
PR				1.0							
	(0.00)	(0.00)	(0.08)		(0.013)	(0.10)	(0.48)	(0.052)	(0.00)	(0.24)	(0.22)
	0.58**	0.59**	-0.54**	0.34**		0.021	-0.034	-0.36**	-0.22*	-0.08	-0.24**
SIZE					1.0						
	(0.00)	(0.00)	(0.00)	(0.00)		(0.81)	(0.71)	(0.00)	(0.011)	(0.34)	(0.00)
	-0.034	-0.017	0.073	-0.19*	0.041		-0.07	0.44**	-0.22*	0.08	-0.13
LEV						1.0					
	(0.71)	(0.85)	(0.43)	(0.050)	(0.65)		(0.44)	(0.00)	(0.012)	(0.37)	(0.15)
	-0.23**	-0.13	0.22*	0.24**	-0.40**	-0.052		-0.098	-0.34**	0.45**	0.35**
LIQ							1.0				
	(0.00)	(0.14)	(0.013)	(0.00)	(0.00)	(0.56)		(0.27)	(0.00)	(0.00)	(0.00)
	-0.066	0.055	0.024	-0.17	-0.16	0.049	-0.17		0.003	0.051	0.015
CF								1.0			
	(0.95)	(0.54)	(0.78)	(0.060)	(2.067)	(0.59)	(0.051)		(0.97)	(0.57)	(0.86)
	-0.069	-0.10	0.049	0.41**	-0.33**	-0.20*	-0.19*	-0.037		-0.13	0.041
AT									1.0		
	(0.44)	(0.25)	(0.59)	(0.00)	(0.00)	(0.024)	(0.035)	(0.68)		(0.15)	(0.65)
	-0.41	-0.002	0.19*	0.13	0.11	0.015	0.003	0.036	-0.036		0.85**
ROA										1.0	
	(0.50)	(0.48)	(0.030)	(0.001)	(0.21)	(0.87)	(0.97)	(0.60)	(0.69)		(0.00)
	-0.14	-0.24**	0.11	0.048	-0.21*	-0.29**	0.11	0.037	0.13	0.25**	
ROE											1.0
	(0.12)	(0.007)	(0.20)	(0.59)	(0.017)	(0.21)	(0.68)	(0.15)	(0.005)	(0.00)	

Pearson correlations are below the diagonal and Spearman correlations are above the diagonal.

P-values are in the parentheses, (**) indicates significance at the 0.01 level and (*) indicates significance at the 0.05 level (two-tailed test). Sample comprises from 17 football clubs participating on the 1st division of the Greek football league from 1994-2004. WIN is the number of wins achieved by the team in every season, PASPOS is a dummy variable taking the value of (1) if the team has finished in the six first positions of the ladder in the previous three periods and zero otherwise, UNCERT is the league's uncertainty measured by the ratio of the points that each team lags from the champion divided by the total points of the champion team, PR is the ratio of earnings to sales, SIZE is the natural logarithm of total assets, LEV is the ratio of total debt to equity, LIQ is the ratio of current assets to current liabilities, CF is the ratio of cash flow to total assets, AT is the asset turnover ration measured by net sales over total assets, ROA is the returns on assets estimated as net income over assets and ROE is return on equity estimated by dividing net income over shareholders' equity.

Pooled Regression Results

The final Table 3 presents the results from the estimation of the profitability equation. In order to control separately the impact of financial and athletic variables on profitability we decomposed the initial equation on two separate equations, where in the first we regress the profitability variable on the WIN, PASPOS and UNCERT variables only, while in the second modification we regress PR on the financial variables. Results were qualitatively the same in all model specifications, thus we limit our analysis on the estimation of the main empirical model (1). All coefficients on both athletic and financial variables have the predicted sign yet only the SIZE, AT, ROA and WIN variables are significant. The coefficient of the

SIZE variable is 1.09 significant at the 0.01 level verifying our assumption that large clubs may be able to generate superior performance since they can exploit economies of scale and organize their activities more efficiently resulting into increased profit streams compared to small clubs. Furthermore, the positive and significant coefficients on both the AT and ROA variables (5.78 and 1.32 both statistically significant) indicates that the ability of football clubs to convert assets into cash can help them to use their resources quickly so as to achieve higher levels of profitability. Finally, among the athletic variables only the WIN variable found positive and significant (0.14) indicating that the short run success of the teams has a significant positive effect on profitability. Specifically, the aforementioned coefficient on the WIN variable (0.14) practically suggests that one win in the season will lead on average, to a 14 per cent increase in the net profit margin. This finding may be proved useful to football clubs managers since the investment on talented players and the improvement of a club's on-field success can enhance their profit making ability.

However, the leverage, cash flow, return on equity and liquidity variables found insignificant. This result can be attributed to the special nature of this specific business organization. Football clubs operate mostly on their human resources (players and trainers) and their non-current assets and consequently this result may be driven by management decisions on their asset al.location on non-current assets, or the minimum importance of financial risk and asset management on profitability.

Table 3: Pooled regression results of profitability equation (1994-2004).

Variables	Model 1	Model 2	Model 3
Intercept	-5.53	-7.67	1.86
SIZE		1.52	1.09
LEV		-0.11	0.10
LIQ		0.52	0.43
CF		0.017	0.003
AT		6.22	5.78
ROA		0.89	1.32
ROE		0.008	0.037
WIN	0.23		0.14
PASPOS	0.18		0.97
UNCERT	1.66		2.21
R²-adj	14.7%	42%	45%

Concluding Remarks

The aim of this paper is to investigate the distinct factors (athletic and financial) that are related to their financial performance of Greek teams. Being more specific, by examining firm specific characteristics such as leverage, liquidity, size, asset turnover, ROA, ROE and cash flows we provide an insight into the football costs that need to be invested and the managerial decisions need to be taken in order to achieve both a prosperous athletic and financial performance.

The results suggest that the profitability of the Greek professional football league is positively affected by the short success of the football clubs but not on the long run success and the uncertainty of the football league. Additionally, the number of wins that a clubs achieve in a season has a significant positive effect on sales suggesting that one win in the season will lead on average, to a 14 per cent increase in the net profit margin. Furthermore, the profitability analysis revealed that large clubs, by means of total assets, may be able to generate superior performance since they can exploit economies of scale and organize their activities more efficiently resulting into increased profit streams compared to small clubs. Finally, our findings suggest that football clubs with increased asset turnover and return on assets have the ability to use their resources quickly and more efficiently so as to achieve higher levels of profitability.

Our findings have implications for the growing body of empirical research on this field, as well as implications for the administrators of the Greek football federation and the managers of the Greek football teams. Specifically team managers can find the results very useful for receiving the proper decisions regarding team's on-field success, in order to improve their financial position.

Regarding future research we must consider additional variables in order to advance the explanatory power of the aforementioned models for instance membership level, stadium capacity, budget expenses etc. Also it will be interesting to examine the issue of the in-house talent development and its impact on the club's accounting disclosure and overall performance, and finally we must consider the on-going debate whether transfer fees paid to football clubs for acquiring players should be capitalized and amortized according to IAS 38.

References

Barbosa, N. and Louri, H., (2005). Corporate Performance: Does Ownership Matter? A Comparison of Foreign-and Domestic-Owned Firms in Greece and Portugal. Review of Industrial Organization. 27: 73–102.

Bird, P. J. (1982). The demand for league football. Applied Economics. 14: 637–649.

Borland, J., and J. Lye, (1992). Attendance at Australian Rules football: a panel study. Applied Economics. 24: 1053–1058.

Borland, J. and R. Macdonald (2003). Demand for sport. Oxford Review of Economic Policy. 19: 478–502.

Cairns, J., Jennett, N. & Sloane, P. J. (1986). The economics of professional team sports: a survey of theory and evidence. Journal of Economic Studies. 13: 1–80.

Burger J.D. and Walters S.J.K. (2003). Market Size, Pay and Performance. A general model application to Major League Baseball. Journal of Sports Economics. 4(2): 108–125.

Copeland, T.E. and Weston, J.F. (1983), "Financial Theory and Corporate Policy." Reading, Mass: Addison-Wesley Publishing Co.

Dobson, S. M., and J. A. Goddard, (1992). The demand for standing and seated viewingaccommodation in the English Football League. Applied Economics. 24: 1155–1163.

Dobson, S. M., and J. A. Goddard, (2001). The Economics of Football. Cambridge University Press

Forrest, D., and R. Simmons (2002). Outcome uncertainty and attendance demand in sport: the case of English soccer. The Statistician. 61: 229–241.

Forrest, D., R. Simmons, and P. Feehan (2002). A spatial cross-sectional analysis of theelasticity of the demand for soccer. Journal of Political Economy. 49: 336–355.

Janssens, P. & Kesenne, S. (1987). Belgian soccer attendances. Tijdschrift voor Economie en Management. 32: 305–315.

Jennett, N. (1984). Attendance, uncertainty of outcome and policy in Scottish LeagueFootball. Scottish Journal of Political Economy. 31: 176–198.

Majundar, Sumit K. (1997). The impact of size and age on firm-level performance: Some Evidence from Indian Industry. Review of Industrial Organization. 12: 231–241

Peel, D. & Thomas, D. (1988). Outcome uncertainty and the demand for football. Scottish Journal of Political Economy. 35: 242–249.

Penman S. (2001). Financial Statement Analysis & Security Valuation. McGraw-Hill

Pinnuck M. and Potter B. (2006). Impact of on-field football success on the off-fieldfinancial performance of AFL football clubs. Accounting and Finance. 46:.499–517

Sloane, P. J. (1969). The labour market in professional football. British Journal of Industrial Relations. 7: 181–199.

Sloane, P. J. (1971) The economics of professional football: the football club as a utility Maximiser. Scottish Journal of Political Economy. 8: 121–146.

Smith, R. & Szymanski, S. (1995). Executive pay and performance, the empirical importance of the participation constraint. International Journal of the Economics of Business. 2: 485–495.

Szymanski, S., and R. Smith (1997). The English football industry: profit, performance and industrial structure. International Review of Applied Economics. 11: 135–153.

Teleworking in United Arab Emirates (UAE): An Empirical Study of Influencing Factors, Facilitators, and Inhibitors

Mohamed G. Aboelmaged and Abdallah M. Elamin

ABSTRACT

This research constitutes an empirical study of influencing factors, facilitators, and inhibitors to the choice of teleworking mode in the UAE context. The research reveals that gender, marital status, nationality, residence location, and work profession are relevant, whereas educational level, Internet use, number of children, age, and years of experience are irrelevant influencing factors for the choice of teleworking mode. Furthermore, the research identifies six distinct facilitators and seven distinct inhibitors. The perceived importance of most identified facilitators and inhibitors to the choice of teleworking mode in the UAE context are found almost similar among the respondents. An exception, however, is made to the association between choice of teleworking mode and individual freedom, travel overload, cost reduction, and union re-sistance. The study outlines the limitations of the present research and suggests some practical implications and recommendations for managers.

Keywords: teleworking, information technology, facilitators, inhibitors, UAE

Introduction

Teleworking has recently received a considerable amount of attention both at the academia and professional world, as one of the remarkable changes in business practices (Morgan, 2004). The last few years have witnessed an increasing interest in the concept of teleworking, particularly in Europe and USA. Current predictions suggest that teleworking may become a common mode of working in future, as Knight (2004) points out that 20 million people in Europe will be teleworking by 2007, taking the enterprise boundary with them.

The concept of using information technology to work at a distance from the regular work site, referred to initially as telecommuting working and later as teleworking. The term first came to wider public attention in the USA in the early 1970s, when it was initially coined by Nilles in 1973 (Nilles, 1994), and it has been described as a growing trend and the future way of organizing work. In some publications, telecommuting and teleworking are often used interchangeably, but telework is generally used in a broader sense, covering a wider array of distributed work. In general, the motives of telecommuting are mainly aimed at achieving travel-time savings, while teleworkers (which may include telecommuters) attempt to work in alternative workplaces.

Literature Review

Various authors (e.g. Mann, 2000) have pointed out the diverse meanings assigned to the term "teleworking." Accordingly, several researchers have tried to establish their own definition. For example Nilles (1994) states that teleworking is — ...the partial or total substitution of telecommunications technologies, possibly with the aid of computers, for the commute to work. In the same vein, Mokhtarian (1991) contends that the term refers to — ...working at home or at an alternate location and communicating with the usual place of work using electronic or other means, instead of physically traveling to a more distant work site. Due to such an inconsistency shaping the definition of the term, one could argue that the definitions applied to telework can be grouped in two main blocks; on the one hand those that emphasize the location of the teleworker and on the other hand, those that stress the use of information communication technologies (ICT).

The empirical literature on teleworking has grown significantly over the last decade and most studies are western-based. Researching teleworking in developing world is unsurprisingly new, an Arab world being no exception. According to Cooper and Schindler (2003), literature can be descriptive, conceptual, empirical, or case study in nature. This section reviews the mainstream empirical teleworking literature.

On empirical side of teleworking research, researchers present results from surveying and analyzing large number of teleworkers, prospected teleworkers, or companies. Golden (2006), for example, use a sample of 393 professional-level teleworkers in one organization to investigate the intervening role of work exhaustion in determining commitment and turnover intentions. Similarly, Neufeld and Fang (2005) conduct two-phased research study to point out that teleworker beliefs and attitudes, and the quality of their social interactions with managers and family members, were strongly associated with productivity. In the similar thought, Thériault et al., (2005) assess differences between home-based working and teleworking behavior among genders and professions considering age groups, household status, car access location within the city and travel distances. They conclude that gender, professional status, and age are influencing factors to the choice to teleworking. For example, older workers are more likely to telework than younger ones, with the exception of lone parents which are seeking for more flexibility. Furthermore, Carnicer et al. (2003) analyze the results of a survey about labor mobility of a sample of 1,182 Spanish employees. Their study indicates that women have lower mobility than men, and that the mobility of men and women is explained by different factors such as employee's perceptions about job satisfaction, pay fairness, and employment stability. In a study of emotional impact of teleworking, Mann (2000) found that respondents of two service industries in the UK perceive teleworking advantages as follows: less travel (57%); more freedom/ flexibility (57%); better working environment (50%); fewer distractions (43%); cheaper (29%); freedom to choose comfortable clothes (14%); freedom from office politics (7%); and easier to complete domestic chores (7%). On the other hand, Mann (2000) found the perceived disadvantages of teleworking include isolation (57%); longer hours (50%); lack of support (28%); less sick leave (21%); career progression (14%); and cost (7%). Similarly, Mannering and Mokhtarian (1995) explored the individual's choice of teleworking frequency as a function of demographic, travel, work, and attitudinal factors. They show that the most important variables in explaining the choice of frequency of teleworking from home were the presence of small children in the household, the number of people in the household, gender of respondent, number of vehicles in the household, whether respondent recently changed departure time for personal reasons, degree of control over scheduling of different job tasks, supervisory status of respondent, the ability to borrow a computer from work if necessary, and a family orientation. In addition, Yap and Tng (1990) conducted a survey of the attitudes of female computer professionals in Singapore towards teleworking. The study reveals that 73% of the 459 respondents were in favor of teleworking. Most would prefer to work at home 1 to 3 days a week and at the office on the other days, rather than working at home full time. They would telework only in times of need (e.g. when they have young children) and were concerned with work and interaction-related problems which might arise from teleworking. Furthermore, Yap and Tng (1990) suggest that teleworking will be of particular interest to employees who are married, those with a high proportion of work that can be done at home, those who find their journey to work frustrating, and those with supervisors and coworkers who are supportive of teleworking.

Research Objectives

The objective of this research is twofold:

1. To examine the influence of specific demographic and individual variables on the choice for teleworking mode.
2. To examine the differences in employees' perception of importance of the facilitators and inhibitors based on their choice of the teleworking mode.

Research Rationale

The rationale behind the study was driven by the fact that most of the teleworking literature has generally taken their roots in the developed countries, most notably North America and Western Europe (Kowalski and Swanson (2005). This point indicates that there is a gap worth filling in the literature resulting from the lack of studies in developing contexts. Considering the uniqueness of the UAE economical, political and socio-cultural contexts, this study would contribute to filling that identified gap.

Though the benefits of teleworking are widely accepted within the literature, there is very scarce empirical research about how demographic and individual variables influence teleworking choice (full-time, part-time, not to telework) in non-western contexts. Examining such relationships between teleworking choice for both actual and prospective teleworker and various demographic and individual variables as well as facilitators and inhibitors in an Arab context, namely UAE will add to the body of knowledge in this regard.

Finally, the outcome of the present study will provide employees, managers and practitioners with important insights that help them make better decisions concerning teleworking programs aiming at improving organizational processes and fostering strategic goals.

Development of Research Hypotheses

The Role of Demographic and Individual Variables

The extant literature has shown that there are numerous demographic and individual variables influence the choice of teleworking mode, including gender, age, martial status, profession, educational level, internet use, nationality, residence, number of children, and years of experience. The subsequent paragraphs review some of the relevant literature on this regards.

Peters et al. (2004) indicate that socio-demographic variables, such as gender and age, are found to influence teleworking adoption and its preference. Similarly, Thériault et al., (2005) suggest that gender and professional status influence teleworking choice, and older workers are more likely to telework than younger ones. Moreover, Yeraguntla and Bhat (2005) show that women households with children are likely to be part-time teleworkers, reinforcing the notion that women are the primary caregivers of children. All in all, they consider age as one of the important individual socio-demographic variable that turned out to be significant predictor of teleworking. The age effect indicates that young adults (less than 25 years) are more likely to prefer part-time employment than older adults. These results are also consistent with the findings of Bagley and Mokhtarian (1997). Moreover, they reveal that race, job type, and length of service are also important influential factors for the choice of teleworking mode. Caucasians and Hispanics, For instance, are more likely to telework than other races (African-Americans, Asians and other). As for job type, their study indicates that employees working for an educational institution are more likely to be part-time teleworkers than employees in other kinds of organizations. For the length of service, Yeraguntla and Bhat's (2005) study reveals that employees who have worked less than a year in the firm are more likely to be part-time teleworkers than those who have been working for longer periods of time.

A survey conducted by Mokhtarian and Salomon (1996) for the employees of the city of San Diego about teleworking, revealed that only 3% of the sample report that they face no constraints to telework but do not have a preference for it and do not currently do it. Based on such a survey they conclude that people who have longer commutes are more likely to report that they want to telework, especially if they are women and younger people. Having children, however, seems to have no effect on the desire to telework.

In the same vein, Mannering and Mokhtarian (1995) use survey data collected from employees of three government agencies in California to model the frequency of teleworking. The results show that being a mother of small children had a positive influence on teleworking, as did the number of vehicles per capita in the household.

Similarly, Wells et al. (2001) conduct surveys of employees at a public agency and a private firm in Minnesota. They find that 43% of the surveyed employees engaged in teleworking. Furthermore, they report that Public agency workers teleworked, on an average, three days a week, while private firm workers teleworked, on an average, 1.92 days a week. The authors find that teleworkers are more likely to be women, married, and have children.

It is worth noting that, Popuri and Bhat (2003) use data from a national survey of 14,441 households conducted by the New York Metropolitan Transportation Council to show factors that increase the likelihood that an individual telework. Such factors include women with children, college education, a driver's license, being married, working part-time, household income, working for a private company (rather than government), and having to pay parking fees at work. Also, it has been found that the longer an individual has worked at her current place of employment, the greater the probability she teleworks.

In their analysis of the telework Survey conducted by the Southern California Association of Governments (SCAG), Safirova and Walls (2004) confirm that having high educational level, more professional experience in general, and a longer tenure with one's current company and one's current supervisor will boost the probability of teleworking. Such a study has also revealed a very surprising finding that teleworkers are more likely to be male and have smaller households than non-teleworkers, which is inconsistent with other studies' findings that have shown women, and especially women with children, to be likely teleworkers.

In the view of the aforementioned discussion, the following hypothesis seems to be relevant for studying the teleworking in the UAE.

Hypothesis 1: There is no difference among employees in their choice for teleworking based on their:

H1a: Gender

H1b: Marital status

H1c: Educational level

H1d: Internet use

H1e: Nationality

H1f: Residence

H1g: No of children

H1h: Age

H1i: Years of experience

H1j: Profession

Facilitators of Teleworking

Teleworking was originally seen as part of a solution to an energy crisis involving the reduction of commuting (Gray et. al., 1993). In this regard, Kurland and Cooper (2002) show that employees choose teleworking to reduce lengthy commutes, to decrease work-related stress, to balance work and family responsibilities, to work longer hours but in more comfortable environments, and to provide uninterrupted time to focus on their work. Organization-wise, teleworking improve employee morale and productivity (Kurland and Bailey, 1999). Interestingly, Gray et al. (1993) find that teleworkers are more productive than office-bound staffs that have to travel to work and tend to suffer a higher level of stress. In addition, Productivity will increase through teleworking if employees are well motivated and satisfied when they are able to manage their own time and assume greater responsibility for their own work. And also because teleworking contributes to the reduction of costs of absenteeism, stress related to traffic congestions, train delays and continuous office interruptions (Lim et al., 2003).

Lupton and Haynes (2000) identify four significant driving forces for teleworking: (1) a change in management attitudes; (2) savings in office costs; (3) demand from staff; and (4) improvements in technology. Other facilitators include improved productivity, improved staff retention, improved morale/motivation, and improved staff recruitment opportunities. These forces are confirmed by Mann (2000) who also points to less travel, more freedom/flexibility, better working environment, fewer distractions, freedom to choose comfortable clothes, freedom from office politics, and easiness to complete domestic chores.

Another classification of teleworking facilitators can be found in the literature is adopted by Mills et al. (2001) and Tung and Turban (1996) who distinguish among three categories of facilitators include organizational, individual, and societal facilitators.

According to Mills et al. (2001) and Tung and Turban (1996) organizational facilitators for teleworking adoption may include securing skilled employees, saving office space, reducing turnover and absenteeism, computer literacy and usage, productivity gains, overcoming limitations of distance and time, providing service from home terminals, and reducing operating cost. Individual facilitators for teleworking, on the other hand, include initiating personal freedom, autonomy, and flexibility (Feldman and Gainey (1997), support no conflicting working environment (Pulido and Lopez, 2005), increasing personal productivity, avoiding a commute, working with fewer interruptions, working in more pleasant surroundings, wearing informal casual clothes, saving the costs of meals, clothes, and commuting, greater time flexibility, greater job satisfaction, and bridging the career gap by avoiding a long career break staying at home (Mills et al., 2001; Tung and Turban, 1996). Community or societal related teleworking facilitators may include reduction of air pollution and dependence on fuel, enable disabled people to work from home, conserve energy and reduce traffic during rush hours and demand on transportation, and solving the problem of rural depopulation (Mills et al., 2001; Tung and Turban, 1996).

Although all these facilitators can support the trend of teleworking implementation, there is still a literature gap about the role of teleworking choice (full-time, part-time, not to telework) in influencing perceived importance of teleworking facilitators. In conjunction with this line of reasoning, the following hypothesis is developed:

Hypothesis 2: There is no difference among employees in the perceived importance of teleworking facilitators based on their choice for teleworking.

Inhibitors of Teleworking

Despite the potential facilitators, teleworking raises two important inhibitors: supervisors' resistance to manage employees that they cannot physically observe (managerial control), and employees' concerns about professional and social isolation (Kurland and Cooper, 2002). Studies, which have addressed these issues, are largely surveys (e.g., Mokhtarian et al., 1995). One exceptional is made to the study conducted by Baruch and Nicholson (1997). They gathered interview data from 62 teleworkers representing five different companies. However, they only noted that isolation and managerial reluctance were factors that could hinder teleworking. In line with this, Reid (1993) cites loss of status and professional isolation as potential dangers for workers moving into teleworking. The likely outcome of isolation is the lack of interaction with colleagues, which stands as a serious inhibitor.

As far as management control is concerned, Kurland and Cooper (2002) has demonstrated that managers may lose control over employees' behavior as employees gain autonomy by teleworking. Teleworking can diminish a manager's perceived control as it physically removes the employee from the conventional work environment. At the same time the employees believe that the isolation may result in lack of promotional opportunities.

Other inhibitors may include cost of implementation and resistance of management to change, longer hours, lack of support, less sick leave, career progression (Lupton and Haynes, 2000; Mann, 2000).

Another classification of teleworking inhibitors is adopted by Mills et al. (2001) and Tung and Turban (1996) who distinguish among three categories of inhibitors include organizational, individual, and societal inhibitors. According to Mills et al. (2001) and Tung and Turban (1996) organizational inhibitors of teleworking adoption may include technology cost inefficiencies, managing out-of-sight employees, need for collaboration with other employees, security risks, problems of supervision, performance control difficulty, work coordination difficulty, legal liability, maintenance of equipment. From the individual point of view, inhibitors may include isolation, doubts and lack of knowledge of the state of a task, unavailability of necessary supplies or equipment, family interruptions and household distractions, no separation of work and home life, lack of interactions with co-workers, and potential lack of loyalty to company, not having a regular routine, workaholics, impedes career opportunities, and missing—what's going on, problem of 'guilt,' and increase in cost of equipment and utilities at home (Mills et al., 2001; Tung and Turban, 1996; Pulido and Lopez, 2005). From the community perspective, teleworking may be inhibited as a result of promoting dispersion of housing, increasing commuting distances, slowing down of real estate market, and declining clothing industry (Mills et al., 2001; Tung and Turban, 1996).

Although all these inhibitors can hinder teleworking implementation, there is a notoriously unfilled literature gap about the role of teleworking choice (full-time, part-time, not to telework) in influencing perceived importance of teleworking inhibitors. Based on the above discussion, the following hypothesis is suggested:

Hypothesis 3: There is no difference among employees in the perceived importance of teleworking inhibitors based on their choice for teleworking.

Research Methodology

This research follows the underlying principles of quantitative research methodology. It entails the collection of numerical data as exhibiting a few of the relationships between theory and research as deductive, and as having an objectivist conception of social reality (Bryman, 2008). A survey research method was applied to obtain insight about the issues explored in the study. Primary research data are collected through structured questionnaire on a voluntary basis. To ensure the right level of teleworking awareness, several studies recommend sampling employees from organizations involved in information technology profession, when studying teleworking (Teo and Lim, 1998; Tung and Turban 1996). The researchers, therefore, consider an employee in an organization within information technology sphere as the unit of analysis in this research. Organizations in Dubai Media City (DMC) and Dubai Internet City (DIC) are selected as target. Both cities include more than 500 organizations in the field of networking, software development, programming, consultancy, broadcasting, publishing, advertising, public relations, research and development, music and creative services. A total of 350 questionnaires are distributed; of these, 148 were returned. 12 questionnaires are ignored due to ignoring complete section(s) or missing data in certain sections, leaving a balance of 136 useful questionnaires for this study, with a valid response rate of 39%. Respondents represent eleven ICT and media organizations specialized in media organization and dissemination, software development, wireless technology, communication tools and equipment, media production, and consultancy services. All organizations are small to medium in size varying from 20 to 300 employees. Questionnaire data were aggregated, and no analysis was conducted linking individual responses to a specific organization.

Measurement Development, Reliability, and Validity

The survey instrument included several statements designed to measure the research constructs. First, choice for teleworking is presented in a nominal scale with three options: (1) not to telework; (2) part-time teleworking; and (3) full-time teleworking. Second, the perceived importance of each of teleworking facilitators and inhibitors is measured based on a four-point Likert scale from "strongly disagree" to "strongly agree." The survey also gathers demographic information on the respondents' gender, marital status, educational level, internet use, nationality, residence location, number of children, age, years of experience, and work profession. A nominal scale is developed for each of these constructs.

Content validity is assessed by examining the process that is used in generating scale items, and its translation into other languages (i.e., Arabic in this study). The determination of content validity is judgmental and can be approached through careful definition of the topic of the concern, the scaled items, and used scales (Cooper and Schindler, 2003). Teleworking facilitators and inhibitors are developed based on extensive review of teleworking literature, and then reduced using a varimax rotated principal component factor analysis. Furthermore, Cooper and Schindler (2003) suggest another way to determine content validity through panel of persons to judge how well the instrument meets the standards. Thus, the researchers conducted independent interviews with two professors of human resources and one professor of information technology applications to evaluate whether research covers relevant constructs. They suggested that the procedure and Arabic translation of the questionnaire were generally appropriate, with some modifications in the translated version of the questionnaire.

Data Presentation and Analysis

Responses from the surveys were coded and entered into SPSS spreadsheets for data analysis. For a descriptive analysis, means, SD, cross tabulation, factor analysis, and Kruskal-Walllis test were applied to the sample.

Profile of Research Demographics

The survey's demographic descriptive statistics are presented in Table 1. Of the 136 respondents, 54.4% select part-time teleworking option, 33.1% decide not to telework, and 12.5% choose full-time teleworking option. 50.7 % of the respondents are male and 49.3 % are female. 67.6% are single and 32.4% are married. 31.7% of married respondents have one child, 26.8% have two children, 22.0% have three children, and 19.5% have four or more children. The research respondents are relatively young; the majority of survey respondents age is between 20 and 29 years (44.9 %), while 25.7% are between 30—39 years, 18.4% are less than 20, and only11% are above 40 years old. The education level reported by respondents showed that 75.7% had university degree or equivalent. Respondents were mainly non-UAE national (66.2%), national Respondents are only represent 33.8%. 40.4% of research respondents live in the emirate of Sharajah 40.4%, Ajman 25.7%, Dubai 22.1%, Abu Dhabi 6.6%, and UmQuin 5.1%. The description shows that 39% of the respondents are internet users for 1-3 times a week, 34.6% use the internet 7 or more times a week, 19.8% use the internet 4-6 times a week, and 6.6% are not using the Internet. According to years of experience, most of the respondents (72.8%) had less than 7 years, and approximately 27.2% had more than 7 years of experience. Respondents in ICT professions are 18.4%, while 27.2% of respondents are in media professions, 27.2% are in management and marketing professions, and 27.2% are in accounting professions.

In conclusion, majority of respondents in this study prefer part-time teleworking, graduate male, single, between 20–29 years of age, care for one child if married, with non UAE nationality, live in Sharjah, use the internet 1-3 times a week, working in different ICT and media professions, with less than 7 years of experience.

Table 1: Profile of research respondents (N=136)

%	N	Marital status
67.6	92	Single
32.4	44	Married
		Nationality
33.8	46	UAE
66.2	90	Non UAE
		Freq. of Internet use
34.6	47	7 or more times /week
19.8	27	4-6 times /week
39.0	53	1-3 times /week
6.6	9	No use /week
		Residence
6.6	9	Abu Dhabi
22.1	30	Dubai
40.4	55	Sharjah
25.7	35	Ajman
5.1	7	UMQ
		Age
18.4	25	Less than 20
44.9	61	20 – 29
25.7	35	30 – 39
11	15	40 or more

%	N	Teleworking choice
12.5	17	Full-time
54.4	74	Part-time
33.1	45	No choice
		Gender
49.3	67	Female
50.7	69	Male
		Educational level
7.4	10	Postgraduate
75.7	103	Graduate
16.9	23	Undergraduate
		Children
31.7	13	1
26.8	11	2
22.0	9	3
19.5	8	4 or more
		Years of experience
36.8	50	0-3
36	49	4-6
17.6	24	7-9
9.6	13	9 or more
		Profession
18.4	25	IT
27.2	37	Media
27.2	37	Mgt. & Marketing
27.2	37	Account. & Finance

Testing the First Hypothesis

A cross tabulation analysis is conducted to assess whether there is no difference among employees in their choice for teleworking based on specific demographic variables. Tables 2 presents frequencies, percentages, and associations of teleworking choice (i.e., full-time, part-time, and not to telework) with a number of selected demographic and individual variables including gender, marital status, educational level, internet use, nationality, residence, number of children, years of experience, and occupation.

Table 2 indicates that there is a significant difference among employees in their teleworking choice based on their gender ($\chi2 = 12.06$, $p < 0.01$). It is clear from the cross tabulation presented in Table 2 that females constitute the majority of employees who select full-time teleworking option (88.2%), while males are the majority who select part-time teleworking (58.1%) as well as not to telework (53.3%). It also shows the association between marital status and teleworking. In that sense, employees' marital status does significantly influence teleworking choice ($\chi2 = 6.69$, $p < 0.05$). The table demonstrates that single employees are over represented among non teleworkers (80%). On the other side, married employees are over represented among full-time teleworkers (52.9%). Educational levels and their distribution cross teleworking choices are illustrated in also reflected in the Table. The analysis suggests no significant difference among employees in their teleworking choice based on their educational level ($\chi2$ = 1.451, n.s). The analysis shows that graduate employees with a university degree or equivalent are over represented in each of teleworking groups; full-time (76.5%), part-time (75.7%), and no teleworking group (75.6%). Similarly, the table suggests no significant difference among employees in their teleworking choice based on their level of Internet use ($\chi2$ = 11.19, n.s.). Employees who use the internet 1-3 times weekly form the majority of two contradictory teleworking groups; full-time teleworking (70.6%) and no teleworking (42.2%). While the majority of employees who prefer part-time teleworking are using the Internet for 7 or more times per week (39.2%). Further, the table indicates that there is a significant difference among employees in their teleworking choice based on their nationality ($\chi2$ = 6.33, p < 0.05). It is clear from the cross tabulation presented in Table 2 that employees with UAE nationality are over represented among full-time teleworkers (58.8%), while employees with non UAE nationality (e.g., Egyptians, Indians, etc.) are over represented among part-time teleworkers (73.0%) as well as non teleworkers (64.4%). Surprisingly, difference among employees in their teleworking choice based on their city of residence is significant ($\chi2$ = 33.99, p > 0.001). Moreover, the table illustrates that part-time teleworking is the main choice of employees living in emirates of Dubai, Sharjah, and Ajman, while the main teleworking choice of employees living in UmQuin emirate is full time. However, employees who are living in Abu Dhabi tend to prefer not to telework. Distribution of number of children cross teleworking choices is also illustrated in the table suggesting that there is no significant difference among employees in their teleworking choice based on their number of children ($\chi2$ = 5.65, n.s.). Employees who select full-time teleworking are equally distributed among those who have two (28.6%), three (28.6%), and four or more (28.6%) children, while part-time teleworking choice is dominated by employees who have one child only (38.5%). Similarly, the table suggests no significant difference among employees in their teleworking choice based on their age ($\chi2$ = 3.78, n.s.). Employees between 20-29 years dominate the majority in every teleworking group; full-time teleworking (47.1%), part-time teleworking (43.2%), and not to telework (46.7%). Moreover, the relationship between employees' teleworking choice and their years of experience is not significant ($\chi2$ = 11.11, n.s.) as demonstrated by the table which indicates that employees who have 4-6 years of experience represent the majority of employees who choose two contradictory options; to telework full-time (56.8%) and not to telework (44.4%), while part-time teleworking choice is dominated by employees who have less than four years of working experience (45.9%). Finally the table illustrates the relationship between teleworking choice and profession. It shows that employees' profession does significantly influence teleworking choice ($\chi2 = 21.95$, p < 0.01). The table demonstrates that 46.7% of employees who prefer not to telework are in accounting and finance profession, 28.4% of employees who prefer part-time teleworking are in management and marketing profession, while employees in media profession are over represented among full-time teleworkers (52.9%).

Table 2: Cross tabulation results

	Teleworking Choice			Total	χ^2	p value
	Full-time	Part-time	No			
Gender					12.06**	0.002
Male	2 (11.8)	43 (58.1)	24 (53.3)	69 (50.7)		
Female	15 (88.2)	31 (41.9)	21 (46.7)	67 (49.3)		
Marital status					6.69*	0.03
Single	8 (47.1)	48 (64.9)	36 (80)	92 (67.6)		
Married	9 (52.9)	26 (35.1)	9 (20)	44 (32.4)		
Educational level					1.43	0.83
Undergrad.	3 (17.6)	11 (14.9)	9 (20)	23 (16.9)		
Graduate	13 (76.5)	56 (75.7)	34 (75.6)	103 (75.7)		
Postgrad.	1 (5.9)	7 (9.5)	2 (4.4)	10 (7.4)		
Internet Use					11.19	0.08
No use	0 (0)	5 (6.8)	4 (8.9)	9 (6.6)		
1-3 times	12 (70.6)	22 (29.7)	19 (42.2)	53 (39)		
4-6 times	2 (11.8)	18 (24.3)	7 (15.6)	27 (19.9)		
7 or more	3 (17.6)	29 (39.2)	15 (33.3)	47 (34.6)		
Nationality					6.33*	0.04
UAE	10 (58.8)	20 (27)	16 (35.6)	46 (33.8)		
Non UAE	7 (41.2)	54 (73)	29 (64.4)	90 (66.2)		
Residence location					33.99**	0.00
Abu Dhabi	0 (0)	2 (2.7)	7 (15.6)	9 (6.6)		
Dubai	2 (11.8)	16 (21.6)	12 (26.7)	30 (22.1)		
Sharjah	5 (29.4)	34 (45.9)	16 (35.6)	55 (40.4)		
Ajman	5 (29.4)	21 (28.4)	9 (20)	35 (25.7)		
UMQ	5 (29.4)	1 (1.4)	1 (2.2)	7 (5.1)		
No. of Children					4.65	0.58
One	1 (14.2)	10 (38.5)	2 (25)	13 (31.7)		
Two	2 (28.6)	8 (30.8)	1 (12.5)	11 (26.8)		
Three	2 (28.6)	5 (19.2)	2 (25)	9 (22)		
Four or more	2 (28.6)	3 (11.15)	3 (37.5)	8 (19.5)		
Age					3.78	0.706
> 20	1 (5.9)	15 (20.3)	9 (20)	25 (18.4)		
20 - 29	8 (47.1)	32 (43.2)	21 (46.7)	61 (44.9)		
30 - 39	6 (35.3)	17 (23)	12 (26.7)	35 (25.7)		
40 ≤	2 (11.8)	10 (13.5)	3 (6.7)	15 (11)		
Years of experience					11.11	0.085
0-3	2 (11.8)	34 (45.9)	14 (31.1)	50 (36.8)		
4-6	10 (56.8)	19 (25.7)	20 (44.4)	49 (36)		
7-9	3 (17.6)	13 (17.6)	8 (17.8)	24 (17.6)		
9 or more	2 (11.8)	8 (10.8)	3 (6.7)	13 (9.6)		
Profession					21.95**	0.001
IT	4 (23.5)	18 (24.3)	3 (6.7)	25 (18.4)		
Media	9 (52.9)	19 (25.7)	9 (20)	37 (27.2)		
Mgt. & Market.	4 (23.5)	21 (28.4)	12 (26.7)	37 (27.2)		
Account. & Finance	0 (0)	16 (21.6)	21 (46.7)	37 (27.2)		
Total	17 (100.0%)	74 (100.0%)	45 (100.0%)	136 (100.0%)		

In conclusion, results from ensuing presentation show significant association between teleworking choice and gender, marital status, nationality, residence, and profession. On the other hand, there is no significant association between teleworking choice and educational level, Internet use, number of children, age, and years of experience. Accordingly, hypothesis H1 is partially supported.

Testing the Second and Third Hypotheses

The data collected concerning employees' perception of teleworking facilitators and inhibitors are reduced using a varimax rotated principal component factor analysis. Tables 3 and 4 display the various facilitators and inhibitors used in this study and show the factor loadings for each of the items. The loadings indicate a significant relationship between items in each of the factors since all but three are greater than .50, the critical value for significant loadings (Hair et al., 1992).

Table 3: Factors analysis of teleworking facilitators

7	6	5	4	3	2	1		
Community concerns (α = 0.83)								
0.093	-0.046	0.202	0.046	0.122	0.097	0.753	Environmental pollution	F20
-0.125	0.235	0.087	-0.055	0.193	0.257	0.749	Working opport. for disabled	F21
0.005	0.169	0.094	0.220	0.008	0.174	0.724	Traffic Jams	F22
0.038	0.088	0.232	0.305	0.141	0.238	0.676	Increasing oil prices	F19
0.169	0.067	-0.059	0.379	-0.020	-0.072	0.647	Severe weather conditions	F24
0.281	-0.082	0.069	0.103	0.320	0.072	0.629	Family care	F23
Professional freedom (α = 0.78)								
-0.076	-0.066	0.125	0.089	0.143	0.782	0.122	Flexible working time and location	F11
0.246	0.214	0.109	0.075	0.190	0.749	0.120	Personal freedom	F9
0.255	-0.041	0.064	0.036	0.146	0.745	0.178	Avoid work stress	F10
0.178	0.108	-0.039	0.368	0.267	0.427	0.152	No absenteeism	F14
Productivity improvement (α = 0.78)								
-0.107	0.167	0.073	0.046	0.785	0.158	0.130	Developing ICT usage	F17
0.109	-0.165	0.221	0.077	0.727	0.007	0.056	Better utilization of working time	F8
-0.018	0.093	0.174	-0.024	0.603	0.439	0.203	Improving output quality and quantity	F13
-0.170	0.284	-0.034	-0.081	0.593	0.419	0.152	Increasing employees loyalty	F16
0.417	0.180	0.176	0.015	0.481	0.237	0.101	Paperless work	F18
Travel load (α = 0.72)								
0.076	0.180	-0.047	0.797	-0.035	-0.126	0.146	Travel preparation	F6
0.036	-0.178	0.132	0.724	0.153	0.247	0.219	Travel time	F5
-0.087	0.049	0.368	0.624	0.007	0.298	0.298	Travel effort and cost	F3
Cost reduction (α = 0.60)								
-0.025	0.017	0.821	0.022	0.180	0.193	0.137	Saving org. space and equipments	F2
0.279	-0.100	0.671	0.136	0.134	-0.003	0.365	Increasing cost of real states	F1
-0.158	0.469	0.518	0.460	0.017	0.102	0.121	Increasing cost of clothes and accessories	F4
Empowering people (α = 0.51)								
0.226	0.801	0.033	0.0091	0.103	0.096	0.102	Minimizing supervisory functions	F15
0.012	0.471	-0.282	0.313	0.355	-0.158	0.217	Task focus	F7
0.831	0.158	0.023	0.042	-0.076	0.209	0.161	Doing other more things	F12
1.382	1.529	1.893	2.346	2.743	2.826	3.512	Eigenvalue	
5.76	6.37	7.89	9.77	11.43	11.78	14.63	Percentage of variance explained	
67.63	61.87	55.50	47.61	37.84	26.41	14.63	Cumulative percentage of total var. explained	
0.993	0.710	0.716	0.743	0.589	0.665	0.645	Standard deviation	

Correlation Matrix Determinant 0.0000179

Kaiser-Meyer-Olkin Measure of Sampling Adequacy = 0.814

Bartlett's Test of Sphericity χ^2 1578.98 df = 276; p = 0.001)

* Principal components analysis; varimax rotation with Kaiser Normalization

Table 4: Factors analysis of teleworking inhibitors

7	6	5	4	3	2	1		
Management concerns (α = 0.73)								
-0.132	0.164	0.055	0.013	-0.201	0.161	0.727	Org. vision and mission are misplaced	B15
0.102	0.060	0.199	0.029	0.224	0.028	0.714	Safety criteria are not guaranteed	B16
0.156	0.147	0.012	0.228	0.045	0.192	0.683	Inapplicable work rules and regulations	B13
0.039	0.261	0.032	0.447	0.088	0.098	0.569	Access difficulty to decision info.	B17
0.265	-0.029	-0.056	0.154	-0.054	0.392	0.559	Hard to control and evaluate performance	B14
Isolation (α = 0.79)								
0.169	-0.015	-0.020	0.124	0.111	0.766	0.089	Misguidance regarding use of org. resources	B11
0.077	0.009	0.257	0.295	0.330	0.672	0.149	Need to interact with work colleagues	B8
-0.167	0.433	0.063	0.199	0.014	0.596	0.265	Lack of teleworking experience	B10
0.205	0.135	0.120	0.062	0.187	0.528	0.385	Feeling guilty toward the organization	B7
0.048	0.293	-0.037	0.480	0.124	0.509	0.194	Missing promotional opportunities at work	B9
Union resistance (α = 0.77)								
-0.207	-0.097	0.139	-0.016	0.800	0.115	-0.045	Union resistance	B23
0.148	0.017	0.027	0.073	0.793	0.104	0.026	Clothing and makeup industry loss	B22
-0.008	0.194	0.036	0.170	0.705	0.196	-0.066	Negative impact on real state sector	B21
-0.010	0.144	-0.020	0.218	0.672	-0.157	0.227	Unclear insurance	B24
Home inadequacy (α = 0.73)								
0.017	0.112	0.192	0.664	0.266	0.053	-0.068	Increased home noise	B20
0.017	-0.091	0.142	0.631	0.150	0.261	0.213	Data insecurity	B19
0.236	-0.148	0.106	0.568	-0.021	0.336	0.438	Coordination difficulty	B12
0.276	0.318	-0.150	0.531	0.020	0.239	0.317	Inapplicable team working	B18
ICT cost (α = 0.80)								
0.134	0.073	0.900	0.017	0.017	0.117	0.130	ICT acquisition cost	B1
0.036	0.145	0.871	0.219	0.125	0.074	0.062	ICT maintenance and upgrading cost	B2
Time mismanagement (α = 0.51)								
0.208	.802	0.068	-0.046	0.021	0.199	0.169	Home time mismanaged	B6
0.094	.594	0.323	0.152	0.289	-0.109	0.091	Org. time expansion	B3
Family intervention (α = 0.63)								
0.835	0.166	0.105	0.217	-0.087	0.111	0.046	Family rights	B4
0.612	0.303	0.187	-0.117	0.095	0.266	0.350	Family – work intervention	B5
1.570	1.747	1.969	2.307	2.595	2.639	3.010	*Eigenvalue*	
6.54	7.28	8.20	9.61	10.81	10.99	12.54	*Percentage of variance explained*	
65.97	59.43	52.15	43.95	34.34	23.53	12.54	*Cumulative percentage of total var. explained*	
0.710	0.694	0.745	0.661	0.688	0.611	0.579	*Standard deviation*	

Correlation Matrix Determinant = 0.00002334

Kaiser-Meyer-Olkin Measure of Sampling Adequacy = 0.794

Bartlett's Test of Sphericity ($\chi 2 = 1345.59$, $df = 276$, $p < 0.001$)

* *Principal components analysis; varimax rotation with Kaiser Normalization*

Cumulative percentage of total variance explained for factor analysis of perceived teleworking facilitators is 67.63% with Kaiser-Meyer-Olkin measure of sampling adequacy = 0.814, while cumulative percentage of total variance explained for factor analysis of perceived teleworking inhibitors is 65.97% with Kaiser-Meyer-Olkin measure of sampling adequacy = 0.794. The Cronbach alpha coefficient is used to assess reliability of the generated facilitators and inhibitors.

As shown in Tables 3 and 4, the alpha reliabilities range from a low of 0.51 to a high of 0.86. All the reliability figures, except two variables, were higher than 0.6, the lowest acceptable limit for Cronbach's alpha suggested by Hair et al. (1992), variables with reliabilities lower than 0.6 deserve a further refinement in future research.

Study-Based Generated Facilitators

The ensuing factor analysis generates six distinct perceived facilitators for tele-working, including community concerns, individual freedom, productivity im-provement, travel load, cost reduction, and empowering people (see Table 3).

(1) Community Concerns

This factor includes a number of community concerns such as reducing envi-ronmental pollution; provision of working opportunities for disabled; reducing traffic jams; continuous increasing of oil prices; severe weather conditions all over the year; and family care issues. No doubt, these concerns make adoption and implementation of teleworking programs in UAE is an appealing option, par-ticularly in case when distance from home to the workplace is far or when traffic congestion is a problem.

(2) Individual Freedom

Items in this factor reflect the notion that teleworking is forced by the need to re-duce stress level and increase job commitment and quality of work life. One likely reason is that the flexibility in working schedule of teleworkers offers opportuni-ties for them to engage in non-work activities to a much larger extent than other-wise possible. Such scheduling freedom may allow time for personal interests.

(3) Productivity Improvement

Items in this factor suggest that improving productivity is perceived as a driving force for teleworking adoption since individuals can avoid interruptions at the of-fice and get work done in an effective and efficient manner. In addition, telework-ing also allows the individual's autonomy by enabling individuals to work during hours where they are most productive (Teo and Lim, 1998).

(4) Travel Load

Items in this factor suggest that the adoption of teleworking will reduce travel burden, including travel preparation time and effort, time of travel, effort con-sumed in the travel, and cost of travel preparation and expenses.

(5) Cost Reduction

Items in this factor reflect the notion that teleworking is a cheap work mode, since it contributes to saving office space and equipments, cost of real states, and cost of clothing and accessories (Mills et al., 2001; Tung and Turban, 1996).

(6) Empowering People

Items in this factor show that teleworking is perceived as a method to empower employees through minimizing supervisory functions and giving the employee opportunity to focus on task at hand.

Study-Based Generated Inhibitors

The ensuing factor analysis generates seven distinct perceived inhibitors for teleworking, including management concerns, isolation, union resistance, home inadequacy, ICT cost, time mismanagement, and family intervention (see Table 4).

(1) Management Concerns

Items in this factor propose that managers may find placing organizational vision and mission, control, supervision, and designing an equitable compensation scheme for teleworker and appraising their performance are difficult (Teo and Lim, 1998).

(2) Isolation

Items in this factor illustrate the concept of professional and physical isolation which is reflected in misguidance regarding use of organizational resources, need to interact with work colleagues, lack of teleworking experience, feeling guilty toward the organization, and missing promotional opportunities at work. This isolation is found to be one of the key inhibitors of teleworking implementation (Kurland and Cooper, 2002; Rognes, 2002).

(3) Union Resistance

This factor reflects the power of union resistance supported by clothing and make-up industry loss, negative impact on real state sector, and unclear insurance. This inhibitor may hinder the implementation of teleworking.

(4) Home Inadequacy

Items of this factor show that home is inadequate place to telework, when teleworkers face increasing home noise, data insecurity at home, work coordination difficulty, and missing the chance of team working.

(5) ICT Cost

Items in this factor suggest that accountability for repairs / maintenance of equipment placed at employees' homes may be a problem. Furthermore, the initial investment in equipment to enable employees to telework may be substantial.

(6) Time Mismanagement

Items in this factor suggest teleworking time is mismanaged and expanded since it intervenes with organization's working time and follows flexible working mode.

(7) Family Intervention

This factor proposes that teleworking may be hindered by the introduction of family rights and family-work intervention process.

Kruskal-Wallis nonparametric test is applied to assess the relationship between employees' perceived teleworking facilitators and inhibitors as ordinal variables and teleworking mode choices as a nominal variable. Results are illustrated in Tables 5 and 6. Table 5 shows Kruskal-Wallis test result of the relationship between perceived teleworking facilitators and teleworking choice. With regard to teleworking facilitators, the analysis demonstrates that there is significant difference among employees in their perceived importance of individual freedom ($\chi 2$ = 17.11, p < 0.01), travel load ($\chi 2$ = 6.76, p < 0.05), and cost reduction ($\chi 2$ = 10.67, p < 0.01) based on their teleworking choice.

On the other hand, there is no significant difference among employees in their perceived importance of community concerns ($\chi 2$ = 5.62, n.s.), productivity improvement ($\chi 2$ = 4.98, n.s.), and empowering people ($\chi 2$ = 4.13, n.s.) based on their teleworking choice. Consequently, hypothesis H2 is partially supported for teleworking facilitators related to individual freedom, travel load, and cost reduction.

Kruskal-Wallis test result of the relationship between perceived teleworking inhibitors and teleworking choice is presented in Table 6 The analysis shows that there is significant difference among employees in their perceived importance of teleworking inhibitors related to union resistance ($\chi 2$ = 6.65, p < 0.01). However, there is no significant difference among employees in their perceived importance of teleworking inhibitors related to all other categories. Accordingly, hypothesis H2 is only supported for teleworking inhibitors related to union resistance.

Table 5: Kruskal-Wallis test result of the relationship between perceived teleworking facilitators and teleworking choice

χ^2	Perceived teleworking facilitators	
	Environmental concerns	
5.54	Environmental pollution	F20
0.52	Working opportunity for disabled	F21
13.28**	Traffic Jams	F22
9.87**	Increasing oil prices	F19
3.45	Severe weather conditions	F24
0.63	Family care	F23
17.11**	*Individual freedom*	
10.77**	Flexible working time and location	F11
9.63**	Personal freedom	F9
13.17**	Avoid work stress	F10
6.34*	No absenteeism	F14
	Productivity improvement	
1.41	Developing ICT usage	F17
0.87	Better utilization of working time	F8
10.15**	Improving output quality and quantity	F13
0.81	Increasing employees loyalty	F16
9.36**	Paperless work	F18
	Travel load	
2.54	Travel preparation	F6
10.33**	Travel time	F5
5.88	Travel effort and cost	F3
	Cost reduction	
15.65**	Saving org. space and equipments	F2
2.85	Increasing cost of real states	F1
3.21	Increasing cost of clothes and accessories	F4
	Supervisory needs	
5.26	Minimizing supervisory functions	F15
0.34	Task focus	F7

*$p < 0.05$, **$p < 0.01$*

Table 6: Kruskal-Wallis test result of the relationship between perceived teleworking inhibitors and teleworking choice

χ^2	Perceived teleworking inhibitors	
	Management concerns	
1.40	Org. vision and mission are misplaced	B15
0.67	Safety criteria are not guaranteed	B16
2.13	Inapplicable work rules and regulations	B13
1.25	Access difficulty to decision info.	B17
2.55	Hard to control and evaluate performance	B14
	Barriers	
3.91	Misguidance regarding use of org. resources	B11
1.76	Need to interact with work colleagues	B8
4.28	Lack of teleworking experience	B10
1.27	Feeling guilty toward the organization	B7
0.31	Missing promotional opportunities at work	B9
	Union resistance	
6.78*	Union resistance	B23
1.78	Clothing and makeup industry loss	B22
0.99	Negative impact on real state sector	B21
9.38**	Unclear insurance	B24
	Home environment	
2.82	Increased home noise	B20
1.80	Data insecurity	B19
1.73	Coordination difficulty	B12
3.79	Inapplicable team working	B18
	ICT cost	
2.57	ICT acquisition cost	B1
1.16	ICT maintenance and upgrading cost	B2
	Time management	
1.83	Home time mismanaged	B6
0.71	Org. time expansion	B3
	Family interaction	
0.11	Family rights	B4
4.16	Family – work intervention	B5

*$p < 0.05$, **$p < 0.01$*

Discussion and Reflection

Demographic Variables

The results have manifested the important role of selected demographic variables in influencing teleworking choice, namely, the role of gender. Accordingly, results of the test have shown that females in the UAE tend to prefer full-time teleworking. Women are found to be motivated by some considerations such as work flexibility, convenience and increased personal freedom (O'Connor, 2001). UAE females have perceived telework as promising avenue to change their traditional work orientation and prove their personal freedom in handling work responsibilities. This is in harmony with Popuri and Bhat (2003), Yap and Tng (1990), and Wells et al. (2001) who suggest that teleworking will be of particular interest to women employees. However, in contradiction to the result generated by this research, some studies show that women employees are not interested in teleworking because they perceived work, not home, as the less stressful and more emotionally rich environment (Hochschild, 1983). In the same thought, Teo and Lim's (1998) study shows that males tend to perceive teleworking as enabling improvement in the quality of life and improvement in productivity/reduction of overheads to a greater extent than females. In line with this argument, Peters et al. (2004) suggest that three out of four teleworkers were male in EU member states, and that this stands in sharp contrast to the widespread opinion that telework was predominantly female. Moreover, research confirms the association between marital status and teleworking choice found in previous research. This is in harmony with Popuri and Bhat (2003), Yap and Tng (1990), and Wells et al. (2001) who suggest that teleworking will be of particular interest to employees who are married. Furthermore, with regard to the educational level, the study indicates no association between educational level and teleworking choice. This result challenges Peters et al., (2004) when mention that well-educated employees were found to be more likely to practice teleworking. Consequently, this research finding is inconsistent with the notion that well-educated individuals are able to telework as they exercise more control over their work schedule than are their co-workers (Yeraguntla and Bhat, 2005).

The present research proves that there is an insignificant association between frequency of Internet use and teleworking choice. Such a finding falsifies the widely held claim that employees master certain level of IT skills including Internet skills are typically suited for teleworking. This research results are inconsistent with the result obtained from the Euro survey 2000, which alleged that telework was most widespread among employees, who used IT frequently in their job (Peters et al., 2004).

Nationality is also found to be significantly associated with teleworking choice. Non-UAE national employees prefer part time and no teleworking compared to UAE national employees who prefer fulltime teleworking. Such findings could be attributed to the fact that Non-UAE national employees attempt to be present at the traditional workplace and establish good work records in order to renew their working contracts, rather than asking for teleworking scheme, though they may prefer. UAE national, on the other hand, are not subject to the stress of being present at the traditional workplace as non UAE employees. This result is in agreement with Yeraguntla and Bhat (2005) who indicate that resident Hispanics are more likely to telework than other races such as immigrants African and Asian who need to demonstrate their working skills, and support their legibility to work and follow work regulations. Similarly, the study shows that residence is associated with teleworking choice. Employees living in Sharjah, Ajman and Umquin are over presented among part-time and full-time teleworkers. This may be interpreted as employees living in these northern emirates always face severe traffic jams in their way to work in Dubai, so that teleworking is perceived as the magic solution for them. This result is consistent with Yen and Mahmassani (1997) when they suggest that the greater the distance from home to workplace, the more likely the employee is to prefer teleworking. Also, Mokhtarian and Salomon (1996) show that people who have longer commutes are more likely to report that they want to telework. This contrasts with Drucker and Khattak (2000) who find that distance to work is negatively correlated with working at home—that is, the farther the individual lives from his job, the less likely he/she to work from home.

Number of children is found to be not associated with teleworking choice. This result confirms Mokhtarian and Salomon (1996) when propose no effect of having children on the desire to telework. Nevertheless, this is in disagreement with Popuri and Bhat (2003), Yap and Tng (1990), Wells et al. (2001) who suggest that teleworking will be of particular interest to employees who have children. Although, working parents may highly value the time-savings of teleworking due to the elimination of commuting time and allow a parent to stay at home with a sick child (Peters et al., 2004), albeit, this is not the case of UAE. In UAE culture, parents (working and not working) depend entirely on foreign maids to take care of their children regardless how many children they may have. Similarly, age is found to be not associated with teleworking choice. In consistent with that, Belanger (1999) does not reveal significant age differences between those practicing telework and those not doing so in her study of a high technology organization in USA. Nevertheless, many studies have revealed contradictory results with regard to the relationship between age and teleworking. Mokhtarian and Salomon (1996), and Bagley and Mokhtarian (1997) show that younger people are more likely to report that they want to telework. Yeraguntla and Bhat (2005) indicate that young adults (less than 25 years) are more likely to be in part-time employment than older adults. This is perhaps a reflection of the fact that many young adults are studying and working part-time at the same time. Inconsistently, the EU member states survey data indicated that the age group 30–49 was over represented among teleworkers (Peters et al., 2004).

Research result related to years of experience tends to be not in agreement with Yeraguntla and Bhat (2005) who suggest that employees who have worked less than a year in the firm are more likely to be part-time teleworkers than those who have been working for longer periods of time. In UAE context, the situation may be different since employees with less working experience try to prove their skills, establish good impression, and get supervisor's support through being presenting at the traditional workplace. After long years of experience, employees may consider teleworking as an alternative work scheme that facilitate managing other concerns such as managing own small business.

The results of the present research prove the existence of an association between employees' profession and teleworking choice. While employees with accounting and finance professions tend to avoid telework, employees with IT, media and management profession tend to telework either on part-time or full-time basis. This result is consistent with Gray et al. (1993) who suggest that computer programmers, systems analysts, catalogue shopping telephone order agents, and data entry clerks fit full-time telework category.

Facilitator and Inhibitors

This research confirms the importance of individual freedom, community concerns, and productivity as key teleworking facilitators perceived by employees. This is in agreement with the mainstream literature that support the perceived importance of personal freedom and autonomy as an immediate symbolic result of employees' interaction with teleworking adoption (Feldman and Gainey (1997; Pulido and Lopez, 2005). In addition, teleworking impact on the society as expressed by employees is clear and tangible on the short run. Mills et al. (2001) and Tung and Turban (1996) consider community and societal related teleworking facilitators such as reduction of air pollution and dependence on fuel, conserve energy housebound and disabled people can work from home, and reduced traffic during rush hours and transportation demand as important determinants of teleworking success in the short run. Moreover, increasing productivity gains is also considered as key derivers for organizations to adopt teleworking (Kurland and Bailey, 1999; Lim et al., 2003; Mills et al., 2001; Tung and Turban, 1996). However, a recent study analyzed the findings of over 80 previous studies, indicating that—little clear evidence exists that telework increases job satisfaction and productivity, as it is often asserted to do (Bailey and Kurland, 2002: p. 383).

As far as teleworking inhibitors are concerned, the present research confirms the importance of isolation as a key inhibitor of teleworking. Recent research indicates that isolation is perceived as one of the key factors that may hinder the implementation of teleworking (Kurland and Cooper, 2002; Rognes, 2002). Isolation is a factor that may result in lack of interaction with colleagues and lack of commitment (Hobbs and Armstrong, 1998). Besides isolation, this research also points to the perceived importance of home inadequacy as a place of working. Although teleworking is treated as working from home, home is perceived by employees as inadequate place for work. Many reasons contribute to this claim involve lack of needed collaboration with other employees, security risks, difficulty of performance control and work coordination (Mills et al., 2001; Tung and Turban, 1996).

Based on the analysis of test results related to hypotheses two and three, most of teleworking facilitators and inhibitors are not associated with teleworking choice. This means that employees with different teleworking modes (i.e., full-time, part-time, not to telework) do not differently perceive the importance of teleworking facilitators and inhibitors. In other words, teleworking facilitators and inhibitors are visible for all employees regardless of their teleworking preference mode. However, teleworking choice is found to be associated with the perception of specific teleworking facilitators and inhibitors. This implies that employees who prefer not to telework tend to perceive less importance for such teleworking facilitator or inhibitor, while employees who prefer to telework part-time or full-time tend to perceive higher importance. Such teleworking facilitators which are associated with teleworking choice include individual freedom, travel overload, and cost reduction. Union resistance is the only teleworking inhibitor that is associated with teleworking choice. This is consistent with other teleworking studies such as Feldman and Gainey (1997) and Pulido and Lopez (2005) who suggest that individual freedom is highly perceived among part-time teleworkers. In addition, teleworkers are more likely to report longer commutes to workplace (Yen and Mahmassani, 1997; Mokhtarian and Salomon, 1996). Finally, perception of cost saving is also over presented among part-time and full time teleworkers in other context (Kurland and Bailey, 1999; Reid, 1993).

Research Limitations

There are several limitations of the present study that may restrict its generalizability. First, sample size is relatively small compared to other studies that have nation-wide samples. Second, the descriptive and exploratory nature of the topic does not allow the researchers to go into the depth of predicting the discovered relationships. Despite that, this study is the first of its kinds to examine teleworking choice and related facilitators and inhibitors in UAE, and in the Arab context. Third, eight out of eleven organizations do not allow the researchers to collect organization's related data such as income of employees, managerial level, degree of computer use in the organization, level of autonomy, decentralization, etc. Consequently, such organization's related variables are eliminated from the original questionnaire in order to maintain access to the respondents. Fourth, as the study focuses on prospective teleworkers in ICT context, results cannot be generalized to other non-ICT contexts.

Practical Implications and Recommendations

The following are some practical implications and recommendations that have emerged from the study of teleworking choice in the UAE:

- Firms employ relatively large percentages of married, female, IT profession, individuals living in remote areas are recommended to adopt flexible work practices such as teleworking.

- Managers are advised to adopt part-time or full-time teleworking scheme in order to integrate and maintain two contradictory strategies; individual freedom and productivity improvement.

- Successful implementation of teleworking requires managers to effectively manage professional and physical isolation of teleworkers through regular office visits and meetings with colleagues. Other practical strategies could include regular e-mail intranet systems, news bulletins and social events. They should also take measures to allow social comparisons to be made, perhaps through use of newsletters, as well as helping teleworkers maintain visibility (perhaps with on-line discussions). Given that these measures are implemented the part-time teleworking is highly recommended compared with full-time teleworking.

- If home is inadequate place for teleworking, managers can rely on telecenters as a substitute. In telecenters, collaboration with other employees can be conducted, and performance control can be facilitated.

- Managers should ensure that any teleworking initiatives are backed up with the appropriate technical support in such way that technicians are available and able to respond quickly to technical problems and equipment failure.

- When initiating teleworking schemes, managers should devise a teleworking policy document that would cover issues such as expectations regarding working when sick, hours to be worked, salary, meetings and visits, deadlines, continuing training, opportunities for career development, management by distance, responsibility for hidden costs (such as electricity) and no hidden costs (such as postage), etc. The aim of such a document is to let workers feel that they have permission to call when they are sick or to switch off the computer at the end of the working day, as well as helping managers manage by outlining to distant workers what is expected of them.

- The importance of data security, privacy, and confidentiality cannot be overlooked when work is performed at home. An organization should invest in the appropriate security measures needed to ensure the confidentiality of data.

- It is necessary to provide training both to the teleworkers and their managers or supervisors. Training areas may include information technologies and networking procedures as well as psychological preparation to work in a new environment.

Conclusion

This study examines the concept of teleworking choice as it applies to UAE context. The relationship between demographic and individual variables, and teleworking choice is investigated. The research reveals that there is no difference among employees in their teleworking choice based on their educational level, Internet use, number of children, age, and years of experience. On the other hand, there is a difference among employees in their teleworking choice based on their gender, marital status, nationality, residence location, and work profession. In addition, the research identifies six distinct teleworking facilitators and seven distinct teleworking inhibitors in the UAE context. Generated facilitators are community concerns, individual freedom, productivity improvement, travel load, cost reduction, and empowering people, while generated inhibitors are management concerns, isolation, union resistance, home inadequacy, information and communication technology (ICT) cost, time mismanagement, and family intervention. Perceived mean importance of these facilitators and inhibitors is computed and ordered. Individual freedom, community concerns, and productivity are perceived by employees as the most important facilitators, while isolation and home inadequacy are perceived as the most important inhibitors. A further statistical test has revealed that there is no difference among employees in the perceived importance of most teleworking facilitators and inhibitors based on their teleworking choice. An exception is the association between teleworking choice and individual freedom, travel overload, cost reduction, and union resistance. The study points out the limitations of the present research and suggests some practical implications and recommendations for managers.

References

Bagley, M.N., and P.L. Mokhtarian (1997), —Analyzing the preference for non-exclusive forms of telecommuting: modeling and policy implications, Transportation, Vol. 24, pp. 203–226.

Bailey, D.E. and Kurland, N.B. (2002), "A review of telework research: Findings, new directions, and lessons for the study of modern work." J. Organizational Behavior, Vol. 23 No. 4, pp. 383–400.

Baruch, Y. and Nicholson, N. (1997), —Home, sweet work: requirements for effective home working, Journal of General Management, Vol. 23 No. 2, pp. 15–30.

Belanger, F. (1999), —Workers' propensity to telecommute: an empirical study, Information & Management, Vol. 35, pp. 139–153.

Bryman, A. (2008), Social Research Methods, Oxford University Press Inc., New York.

Carnicer, M., A. Sanchez, M. Perez, and M. Jimenez (2003), —Gender differences of mobility: analysis of job and work-family factors, Women in Management Review, Vol. 18 No. 4, pp. 199–219 Cooper, D and Schindler, A. (2003), Business Research Methods. Irwin, MA.

Drucker, J., and A.J. Khattak (2000), —Propensity to work from home—modeling results from the 1995 nationwide personal transportation survey, Transportation Research Record, Vol. 1706, pp. 108–117.

Feldman, D. and Gainey, T. (1997), —Patterns of telecommuting and their consequences: framing the research agenda, Human Resource Management Review, Vol. 7 No. 4, pp. 369–88.

Gray, M., Hodson, N. and Gordon, G. (1993), Teleworking Explained, John Wiley & Sons, New York, NY.

Golden, T. (2006), —Avoiding depletion in virtual work: Telework and the intervening impact of work exhaustion on commitment and turnover intentions, Journal of Vocational Behavior, Vol. 69, pp. 176–187.

Hair, Joseph, Ralph E. Anderson, and Ronald L. Tatham (1992), Multivariate Data Analysis with Readings, 3rd edition, Macmillan Publishing Company, New York.

Hobbs, D. and Armstrong, J. (1998), —An experimental study of social and psychological aspects of teleworking, Industrial Management & Data Systems, Vol. 98 No. 5, pp. 214–8.

Hochschild, A. (1983), The Managed Heart: Commercialization of Human Feeling, University of California Press, Berkeley, CA.

Knight, William (2004), —Working drives switch to federated access rights, Info Security Today, September/October, p. 22–25.

Kowalski, K. and J. Swanson (2005), —Critical success factors in developing teleworking programs, Benchmarking: An International Journal, Vol. 12 No. 3, pp. 236–249.

Kurland, N. and Bailey, D. (1999). When workers are here, there, and everywhere: a discussion of the advantages and challenges of telework. Organizational Dynamics, pp. 53–68.

Kurland, N. and Cooper, C. (2002), —Manager control and employee isolation in telecommuting environments, Journal of High Technology Management Research, Vol.13, pp. 107–126.

Lim, H., A. van der Hoorn, V. Marchau, (2003), —the effects of telework on organization and business travel, Paper submitted for —Symposium on Teleworking, 4th Interbalkan Forum International IT conference, Sofia, Bulgaria, 6-7 October 2003.

Lupton, P. and Haynes, B. (2000), —Teleworking—the perception-reality gap, Facilities, Vol. 18. No. 7/8, pp. 323–8.

Mann, Sandi (2000), —An exploration of the emotional impact of tele-working via computer mediated communication, Journal of Managerial Psychology, Vol. 15 No. 7, pp. 668–690.

Mannering J. and P. Mokhtarian (1995), —Modeling the choice of telecommuting frequency in California: An exploratory analysis, Technological Forecasting and Social Change, Vol. 49, pp. 49–73.

Mills, J., Wong-Ellison, C., Werner, W., and Clay, J. (2001), —Employer liability for telecommuting employees, Cornell Hotel and Restaurant Administration Quarterly, Vol. Oct-Nov., pp. 48–59.

Mokhtarian, P. L. (1991), —Definig Telecommuting, Transportation Research Record, Vol. 1305, pp. 273–281.

Mokhtarian, P. L., Handy, S. L. & Salomon, I. (1995), —Methodological issues in the estimation of the travel, energy, and air quality impacts of telecommuting, Transportation Research, Vol. 29, pp. 283–302.

Mokhtarian, P. L., and I. Salomon (1996) Modeling the choice of telecommuting: the importance of attitudinal factors in behavioral models. Environment and Planning, Vol. 28, pp. 1877–1894.

Morgan, Robert (2004), —Teleworking: an assessment of the benefits and challenges, European Business Review, Vol. 16 No. 4, pp. 344–357.

Neufeld, D. and Fang, Y. (2005), Individual, social and situational determinants of telecommuter productivity, Information & Management, Vol. 42, pp. 1037–1049.

Nilles, J. M. (1994), Making Telecommuting Happen: A Guide for Telemanagers and Telecommuters. Van Nostrand Reinhold, New York.

O'Connor, V. (2001), —Women and men in senior management, Women in Management Review, Vol. 16 No. 8, pp. 400–4.

Peters, P., Tijdens, K. and Wetzels, C. (2004), —Employees' opportunities, preferences, and practices in telecommuting adoption, Information & Management, Vol. 41 No. 4, pp. 469–82.

Popuri, Yasasvi, and Chandra R. Bhat. 2003. —On modeling choice and frequency of home-based telecommuting, Transportation Research Record, Vol.1858, pp. 55–60.

Pulido, J., and Lopez, F. (2005) —Teleworking in the information sector in Spain, International Journal of Information Management, Vol. 25, pp. 229–239.

Reid, A. (1993), Teleworking as a Guide to Good Practice, NCC Blackwell.

Rognes, J (2002), —Telecommuting resistance, soft but strong: Development of telecommuting over time, and related rhetoric, in three organizations, SSE/EFI Working Paper Series in Business Administration, No 2002:1, Stockholm School of Economics, Sweden.

Safirova, Elena and M. Walls, 2004. "What have we learned from a recent survey of teleworkers? Evaluating the 2002 SCAG Survey." Discussion Paper 04–43. NW: Resources for the Future, Washington, D.C.

Teo, T. and Lim, V. (1998), —Factorial dimensions and differential effects of gender on perceptions of Teleworking, Women in Management Review, Vol. 13 No.7, pp. 253–263.

Thériault, M., P. Villeneuve, M. Vandersmissen, and F. Des Rosiers (2005), Homeworking, telecommuting and journey to workplaces: are differences among genders and professions varying in space?, the 45th. Congress of the European Regional Science Association, 23–27 August 2005, Vrije Universiteit Amsterdam.

Tung, L-L. and Turban, E. (1996), —Information technology as an enabler of telecommuting, International Journal of Information Management, Vol. 16 No. 2, pp. 103–18.

Wells, Kimberly, F. Douma, H. Loimer, L. Olson, and C. Pansing. (2001), —Telecommuting implications for travel behavior: case studies from Minnesota. Transportation Research Record, Vol. 1752, pp. 148–5.

Yap, C. S. and Tng, H. (1990), —Factors associated with attitudes towards telecommuting,' Information & Management, Vol. 19, pp. 227–235.

Yen, Jin-Ru, and Hani S. Mahmassani. (1997), —Telecommuting adoption: conceptual framework and model estimation. Transportation Research Record, Vol. 1606, pp. 95–102.

Yeraguntla, A. and C. Bhat (2005), —A Classification Taxonomy and Empirical Analysis of Work Arrangements, Working Paper # 05-1522, The University of Texas, Austin.

The Role that Personality and Motivation Play in Consumer Behaviour: A Case Study on HSBC

Jolene Montgomery

Introduction

In today's information-oriented society, research and development, particularly information research, has become an important activity for business companies, institutions, and organizations, who want to know more about the consumer market, people who consider consumption as embedded and part of their everyday lives. Initially, commercialism of goods and services through advertising mainly focused on extant products and services that people need; nowadays, persuasive messages are extended through advertising, informing people about the goods and services that they should and ought to know and buy for themselves.

Selling these products and services through persuasive advertising messages, however, are the products of advertising research. More specifically, consumer research tries to identify not only the socio-demographic, but also psychographic profile of consumers, understanding how people can be persuaded to buy a company's product or service. Consumer research looks into the motivations and personalities of an individual in terms of consuming or buying a particular product or service, later turning this information into strategies geared at gaining a particular segment of the market that the company targets or centers on.

This paper discusses in detail the role that motivation and personality plays in influencing consumer behavior, taking the case of the Hongkong and Shanghai Banking Corporation (HSBC) as an example to discuss and analyze these important points. In this paper, an analysis of the print ads of HSBC is analyzed, relating its features to identify its target market and perceived motivations and personalities of HSBC's target market. This study aims to provide an illustration of how motivation and personality analysis of consumers are vital to the understanding of consumer markets and behavior.

Consumer Personality and Motivation as Illustrated in HSBC Print Ads

The terms motivation and personality may seem familiar for people, but its significance to consumer behavior is less known, yet increasingly essential in identifying, determining, and understanding insights regarding consumption patterns and preferences.

Personality is defined by Sheth et. al. (1999) as "[a] person's consistent ways [sic] of responding to the environment in which he or she lives" (G-11). Personality, he states, is created through the combining of external influences or the social environment and genetic or biological traits of the individual. The combination of social with the individual results to the creation or development customer personality; consumer personality may be product- or service-oriented, or both (243). Product-oriented consumers tend to patronize a product or service based on the merchandise itself, while service-oriented consumers tend to "seek relationships" with the seller, producer of the service or manufacturer of the product.

Motivation, meanwhile, is identified as "an inner drive that reflects goal-oriented arousal" (Arnould et. al., 2004:259). It differs from personality in that it is a deeper and more abstract concept, although similar to it in the sense that motivations are also linked to the social environment and individual traits of the individual.

In order to distinguish properly between the two terminologies, personality may be understood as a holistic or general term to describe consumer behavior—that is, consumer behavior at the macrolevel. Motivation, on the other hand, provides an in-depth look of the consumer as a unique individual and harder to discern and understand. Motivation, thus, represents consumer behavior at the microlevel.

Applying these concepts in the context of HSBC and its adverting and marketing strategies, the researcher of this paper analyzed three (3) print ads, which came out on three issues of TIME magazine: March 11, 18, and 25 for the year 2002. These issues are relevant for this study, since this is the year that HSBC launched its campaign on multiculturalism, entitled "Business Connections," carrying with it the slogan, "Never underestimate the importance of local knowledge."

These ads, which serve as units of analysis of the study, are textually analyzed to generate themes that depict the influence of motivation and personality of the target consumer market in the production of the advertising message and form. The texts that follow discusses the salient points generated from the analysis, which involves the following: (1) HSBC ads illustrate how culture as the social environment and cultural traits serve as primary consideration in crafting today's advertising messages to the consumer; (2) motivations such as achievement, power, uniqueness, affiliation, and self-esteem are the determinants of a consumer's cultural environment and traits. These linkages are seen thoroughly in the next section.

"Never Underestimate the Importance of Local Knowledge": HSBC on Culture and the Consumer

It was initially discussed that consumer behavior in terms of the personality dimension is primarily determined through the consumer's merchandise- or service-orientedness. However, going further into the analysis of consumer personality, there is also recognition that consumers adopt various personalities depending on the role that they assume as consumer: user, payer, or buyer (Sheth et. al., 1999:243).

These distinctions are somewhat similar to the merchandise-oriented and service-oriented dichotomy introduced earlier, although in this new set of consumer personalities, the economic dimension is taken into consideration. Thus, consumer personality is divided into three facets: the merchandise-oriented consumer as the user, the service-oriented consumer as buyer, and the consumer's financial position as the payer.

In the HSBC "Business Connections" advertising campaign, these components of consumer behavior become evident. The buyer or service-oriented consumer, clearly, is the primary target market of HSBC, since its business orientation is to provide banking services for potential clients. However, it is evident that HSBC decided to transgress its role as bank service provider to becoming the consumer's "expert" on cultural knowledge of every nation engaged in the business industry. Its ad campaign is developed to fulfill this objective, allowing the consumers to get to know other societies and cultures around the world, stressing how HSBC knows each culture featured, and understands these cultures well, enabling them to effectively handle business and financial transactions with them.

The March 18, 2002 ad of HSBC shows two images of hands toasting glasses together, with the first image displaying the text, "HUNGARY: Bad luck," while the second image shows the text "USA: Good health." This simple, yet effective display of cultural differences around the world is reflective of HSBC's work ethic, implying to the reader (who is also a potential customer) that they understand various cultures very well. This understanding of various cultures of the world becomes HSBC's advantage, since it helps them "...recognise [sic] financial opportunities invisible to outsiders" (TIME, 2002).

The buyer (service-oriented consumer) would be enticed with the said ad, since institutions whose primary commodity is service provision knows that understanding the individual and his/her social/cultural environment is one way of highlighting the fact that they are selling service that reflects the personality of consumer-buyer. For example, an American who encountered the HSBC ad discussed above will show approval of the distinction made by the banking company. This approval will, in turn, lead to the development of credibility of HSBC and trust from the consumer; thus, the next time the consumer should consider seeking help in banking services of financial transactions, s/he would seek HSBC, primarily because it understands the individual's sentiments, feelings, and opinions—in other words, the customer's personality.

This can also be applied to the consumer as user, wherein the commodity can be either the information about Hungarian and American culture depicted in the ad, or the banking service provided by HSBC, or both. The first scenario occurs when the individual is also a business person who is interested in knowing the business culture of another country, which, to the business-oriented individual, is always a potential customer. The information that it is bad luck in Hungary and good health in the USA to toast glasses will become helpful to the individual when s/he deals with clients coming from these countries. Or, it may be that the consumer-user is interested in knowing up to what extent HSBC's credibility and banking experience is; thus, from the ad, one can already surmise that HSBC has sufficient experience to merit credibility and warrant trust from the consumer.

The consumer as payer, meanwhile, is different from the preceding kinds of consumer personalities discussed. Since in this category, the impact of economic stability and financial capability are taken into account, the strategy of HSBC must then emanate as to include and address the concerns of the consumer-payer in its ad. In the ad, this concern is addressed by the inclusion of the term "local knowledge" in its ad, giving reference to the fact that not only did HSBC learned to know and understand the culture of a country, but it was also able to capture its people, the masses, which ultimately constitutes a particular culture, the keepers and actors of this "local knowledge."

Whether the consumer-payer has financial troubles or not, HSBC considers everyone a potential customer, and the consumer-payer, through the ad, shall assume that HSBC will aid him/her for whatever banking service or financial support s/he wants to avail, simply because the institution understands "local knowledge," the reality of life in that particular society in the world. HSBC, then 'personalizes' its service, accommodating for any diversity that it may encounter in the conduct of its business (Francese, 2004:40-1).

In effect, business institutions like HSBC have learned to ascertain the kinds of people who can be their potential clients or customer by categorizing them according to their cultural and individual traits. In this section, personality is shown to be a major determinant in identifying the likelihood of an individual to become a consumer of a particular product or service.

Thus, if the consumer is merchandise-oriented or a user, then goods as commodities are more appealing for him/her; the consumer as buyer, on the other hand, is more interested in establishing close ties with the sellers, producers, or manufacturers, the people who provide service for the consumers; and lastly, the consumer as payer takes into consideration the economic climate of the institution itself, and aligns it with his/her own financial status, and when both criteria meet, it is then that the consumer-payer decides whether to subsist to the product (or service) or not.

Motivations as the Consumer "Drive": Achievement, Power, Uniqueness, Affiliation, and Self-Esteem

In the previous section, the researcher has identified consumer personality in terms of the product, service, and financial climate of both the seller and buyer. In this section, the study takes an in-depth look at the individual traits of these consumers, identifying potential motivations that serve as consumer 'drives,' needs and wants that may have been accomplished through the consumption of the product or service offered by the business company/institution. In this section, motives are identified into four, namely, achievement, power, uniqueness/novelty, affiliation, and self-esteem motive. These kinds of motives are present in one way or other within the consumer as s/he goes through the decision-making process (of purchasing a product, good, or service). These kinds of motives are defined by Arnould et. al. (2004) as follows:

- Achievement motive—the drive to experience emotion in connection with evaluated performance;
- Power motive— the drive to have control or influence over another person, group, or the world at large;
- Uniqueness/novelty motive—the drive to perceive oneself as different from others;
- Affiliation motive—the drive to be with people; consumers sometimes experience a strong motivation to reconnect and associate with groups…; and
- Self-esteem motive—credit for successes, explain away failures, (consumers) see themselves as better than most others.

Given the following kinds of motivations and their definitions, the HSBC advertising campaign clearly invokes all of these motivations. This is actually imperative for the company, since they have as their audience people of all ages, gender, races, and culture. In order to become effective in motivating and persuading consumers to subsist to the HSBC service, the company must be able to "grasp" or capture all of the characteristics of its consumers; thus, the need to consider all the motivations that a consumer may have or use in the process of deciding to purchase a product or subscribe to a particular service.

The "Business Connections" ad campaign illustrates the achievement motive, primarily because HSBC tries to establish personal relations with the consumer. In trying to sell its service (banking service), HSBC evokes the emotions of consumers as it re-establish ties with him/her by showcasing a particular culture of a nation (Benady, 2004:43). For example, the HSBC ad depicts the culture of UK and US by illustrating an image of cross-legged shoes, where the act may be construed as "relaxed" in American culture, while it is considered "rude" among Thai people. Feelings of approval and/or reproach over the said cultural norms shows how achievement—that is, to achieve and conform with the cultural norm—becomes a primary factor that convinces the consumer to subsist to the product or service. Similarly, the illustration of a particular culture through print advertisement creates the impression that every culture has power, thereby influencing the consumer that as member of that culture, s/he has power to approve or disapprove of the ad and patronize or not patronize the product or service advertised.

The most important and dominant motives included in the ad are the uniqueness, affiliation, and self-esteem motives. In the HSBC ad campaign, the showcasing of a particular culture per ad illustrates the strong regard the company has for cultures of the world—that is, the uniqueness of a society or nation and the individual. These motives work in linkages, initially starting with starting with the premise that every culture is unique, and proof of this in the ad is the provision of information telling the readers the uniqueness of a particular material, symbol, or activity in the business culture of a particular country (Brown et. al., 2003:19).

In the same way, affiliation results from the character of uniqueness, wherein the consumer, once s/he has identified herself/himself with the culture depicted in the ad, feels a sense of belongingness with the subject or image displayed in the ad. Thus, the consumer who understands and agrees with the culture depicted becomes affiliated with it, and HSBC, taking advantage of this event (creation of uniqueness and sense of affiliation), persuades the reader to subsist to HSBC, the company that 'understands' people of all cultures. Combining both uniqueness and affiliation results to the development of self-esteem, where the consumer, once s/he recognizes the value that HSBC gives to people of his/her culture, would then affiliate himself/herself with the banking institution. In effect, the ad portrays the HSBC brand as embedded in each culture and actually belongs to the people. Thus, the ad campaign becomes successful in showing the consumers that HSBC knows its consumers—their personality, and what motivates them to consume a particular product or service.

Conclusion: Motivations and Personality Reflects the HSBC Consumer

In the previous sections of this study, motivation and personality were shown as essential factors identifying and reflecting consumer behavior. It is evident that both sociodemographic and psychographic characteristics both determine consumer personality and motivations (Rafee, 2004:17).

Personality primarily determines the consumer as buyer, payer, and user of a product and/or service. In so doing, the sociodemographic variable income and socio-economic class helps determine the extent up to which the consumer shall adopt the roles enumerated earlier. Furthermore, attitudes towards a product or service—that is, becoming product- or merchandise-oriented or service-oriented as a consumer is a psychographic variable that also aids him/her to buy and consume the product/service. Both variables work together to ultimately determine the consumer's choice and preference of a product or service.

Motivations are also important determinants of the consumer's characteristics and behavior towards consumption. Along with sociodemographic variables of income, age, sex, and ethnic membership, psychographic variables such as need for achievement, affiliation, power, and self-esteem are also taken into account. Combining these variables together result to the fulfillment or failure of the consumer to achieve his/her needs, depending on the quality of the product or service offered—that is, up to what extend consumers' needs are achieved through the product/service.

It is important to provide a case in point to illustrate how consumer characteristics (personality and motives) eventually influence consumer behavior. In the case of the HSBC ad campaign, we have seen how personality and motivations become components of the ads, allowing it to create and transmit an effective advertising message to the reader/consumer. Centering on the issue of cultural sensitivity and tolerance to people's and individual's differences become the vehicles of persuasion that HSBC uses to illustrate the personality traits and motivations of the consumer. In the case of HSBC, research on the profile of the HSBC consumer showed that there is indeed a large percentage of today's consumer market who came from various ethnic or cultural origins apart from the American (or Western) consumer (HSBC CSR Report, 2003).

In sum, this study has shown that taken to its advantage, research and information concerning consumer behavior shows that two of its influential factors are personality and motivation. The future of better and more effective advertising and marketing promotions of the consumer market can thus be further improved, as was shown in the HSBC experience, wherein consumer personality and motivation are taken into consideration and acknowledged as imperative in determining the consumer market in today's information age.

Bibliography

Arnould, E., l. Price, and G. Zinkhan. (2004). Consumers. Boston: McGraw-Hill/Irwin.

Benady, D. (June 2004). "It's time to get personal." Marketing Week.

Brown, S., R. Kozinets, and J. Sherry. (July 2003). Teaching old brands new tricks: Retro branding and the revival of brand meaning. Journal of Marketing, Vol. 67.

HSBC Corporate Social Responsibility Report 2003.

Rafee, A. (May 2004). "The key to targeting lies in users' real behaviour." New Media and Age.

Sheth, J., B. Mittal, and B. Newman. (1999). Consumer Behavior and Beyond. NY: Harcourt Brace.

Mobile Technology and the Value Chain: Participants, Activities and Value Creation

Constantinos Coursaris, Khaled Hassanein and Milena Head

ABSTRACT

Technology has evolved significantly and it is increasingly being used by businesses and consumers alike. Technologies such as those supporting electronic business (e-Business) and mobile business (m-Business) are being used across organizations extensively in an attempt to improve operations and subsequently translate in either financial gains or strategic advantages. Opportunities for realizing either of the two types of benefits can be identified through an examination of a business' value chain.

This conceptual study begins by proposing a business-centric interaction model that helps explain the interactions among all participants involved in an organization's possible activities. The paper then explores the potential fit of wireless and mobile technologies across a company's value chain through the citation of potential mobile and wireless business applications currently available. Finally, a discussion on the expected benefits and relevant concerns of mobile technology, as well as considerations for future research are provided.

Keywords: mobile technology, value chain, mobile applications, m-Business, concerns

Introduction

Technology has evolved significantly and it is increasingly being used by businesses and consumers alike. For businesses, the last two decades have been marked by the transition of large and cumbersome mainframe computing systems, to personal computers offering increased capabilities and occupying only a small area of personal and work space. The latest innovation is found in mobile devices that introduce higher levels of flexibility and personalization. Technologies such as those supporting electronic business (e-Business) and mobile business (m-Business) are being used across organizations extensively in an attempt to improve operations and subsequently translate in either financial gains or strategic advantages. Opportunities for realizing either of the two types of benefits can be identified through an examination of a business' value chain.

The paper begins by defining m-Business and presenting a business-centric interaction model that helps explain the interactions among all participants involved in an organization's possible activities. Then, an overview of the value chain and the impact of m-Business on it are provided through the citation of potential mobile and wireless business applications currently available. Finally, a discussion on the expected benefits and relevant concerns of mobile technology, as well as considerations for future research are provided.

M-Business

Mobile business (m-Business) can be defined as electronic business interactions/transactions enabled at least in part by mobile technology that may target businesses and consumers alike (Coursaris and Hassanein, 2002). For the purpose of this paper the term m-Business incorporates m-Commerce activities which represent the transactions enabled by mobile technology.

There are several mobile technologies that support m-Business. These are typically grouped as devices and networks (White, 2005). Mobile devices range from small radio frequency identification (RFID) and global positioning system (GPS)

chips to barcode scanners and wirelessly-enabled handheld personal computers. Mobile networks range from Bluetooth and RFID readers to mobile telecommunications networks and GPS. These mobile technologies are being used by organizations to help address their needs while offering opportunities for flexibility and customization.

Unlike e-Business, which leverages wired and consequently immobile access points (e.g. PCs), m-Business offers value by enabling users to be mobile and reachable anytime and anywhere. Therefore, value creation can occur by supporting either mobile users (e.g. employees) or mobile activities (e.g. tracking raw materials and supplies). A growing industry trend is found in Fixed-Mobile Convergence (FMC), in which centralized management and infrastructure support a mobile workforce, providing —full access to business applications from any location or network connection (Winther, 2007). Thus, the greater the size of the mobile workforce and/or the higher the ratio of mobile activities within an organization, the greater the value proposition of m-Business for a firm. It is therefore important to explore the types of wireless interactions relevant to businesses.

A Business-Centric Model of Mobile Interactions

In crafting the value proposition of m-Business for a business, three components are of interest: relevant actors, unique attributes of mobile technology, and the types of activities supported. We begin by identifying the relevant actors. These are described below and included in Figure 1, where interactions occurring among them within a wireless environment (i.e. at least one actor is using the wireless channel) are mapped:

Employees (E)

These are individuals that are part of an organization (in Figure 1 the association is identified by the matching subscripts, e.g. Business 1 has two employees E1A and E1B). Employees may need or want to interact with other colleagues or employees of other businesses. In addition, employees may be at the receiving end of an interaction initiated by both internal and external information systems. One example of a business application in this area is wireless notification by a System via SMS for a critical update. To this end, the possible wireless interactions are Employee-to-Employee (E2E), Employee-to-Consumer (E2C), and Employee-to-System (E2S). It is important to note that most such interactions could naturally involve activities in the reverse direction, e.g. a wireless System-to-Employee (S2E) interaction mode as well.

Systems (S)

These are machines that are run by businesses and could either be front-end (e.g. web interface) or back-end systems (e.g. corporate database). An example of this type of interaction is an employee engaged in wireless (and possibly remote) access of the business' Enterprise Resource Planning (ERP) system. To this end, the potential wireless interactions are System-to-Consumer (S2C), System-to-Employee (S2E), and System-to-System (S2S). Again, the activity could occur in the reverse direction as well.

Consumers (C)

These are individuals that a business may interact with wirelessly. One example is an interaction between an employee and the consumer by means of SMS or e-mail. To this end, the potential wireless interactions are Consumer-to-System (C2S), Consumer-to-Employee (C2E), and Consumer-to-Consumer (C2C) to the extent it relates to the business activities (e.g. community-based interactions).

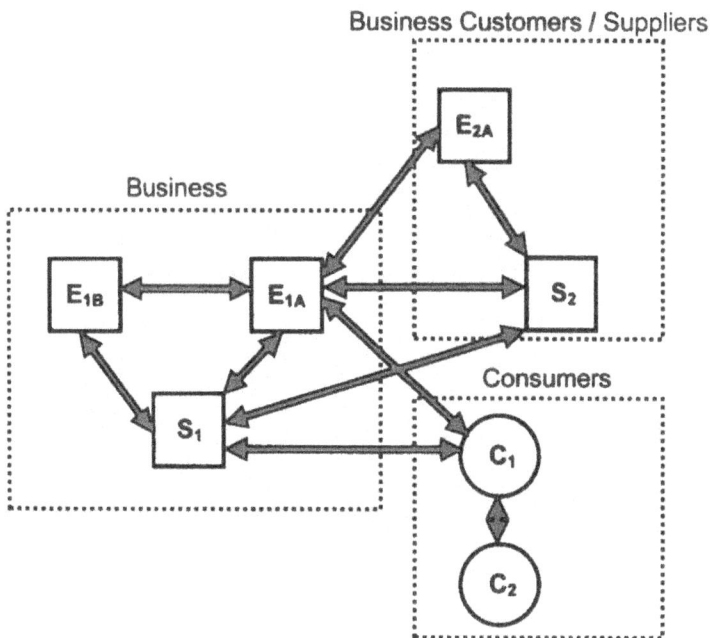

Key:
The Business entity shows two potential employees (E_{1A} and E_{1B}), and a potential internal I.T. system (S_1)
The Business Customers and/or Suppliers entity shows a potential employee (E_{2A}), and an I.T. system (S_2)
The Consumers entity shows two potential consumers (C_1 and C_2)

Figure 1: A Business-Centric Model of Mobile Interactions

Having identified the mobile interactions, the next relevant component in formulating a value proposition for mobile technology to organizations is to understand its unique or enhanced attributes, which include connectivity, personalization, and localization (Turban, 2002).

Connectivity

A wireless infrastructure enables mobile workers with 24/7 connectivity supporting "anytime, anywhere" communication and information exchange.

Personalization

Mobile devices are typically assigned to single users, who are then able to personalize interface and application settings that may not only increase their satisfaction with using the device but may also improve the efficiency and effectiveness of the system.

Localization

Localization is particularly important as it adds a new dimension to reachability extending from the Internet's ability to reach a location (i.e. IP address) to reaching a user (i.e. a mobile worker) or an item (e.g. tracking a shipment).

The context of value creation for mobile technology becomes complete by the types of organizational activities supported. These activities are explored next in more detail within the framework of Porter's (1985) value chain.

The Value Chain

Michael Porter (1985) coined the term value chain as the set of linked activities performed by an organization that impact its competitiveness. As seen in Figure 2, the value chain consists of five primary and four support activities. Primary activities are directly concerned with the creation or delivery of a product or service. These include inbound logistics (e.g. receiving and storing raw materials), operations (e.g. converting raw materials through manufacturing into finished goods or service creation process), outbound logistics (e.g. delivering of goods or services to customer), marketing and sales (e.g. identifying opportunities and processing customer orders) and service (e.g. providing after-sales support to customers). These primary activities are facilitated by support activities, which include infrastructure (e.g. organization-wide administrative and managerial systems), human resource management (e.g. managing personnel), technology development (e.g. R&D and continuous enhancements of technology-related activities), and procurement (e.g. purchasing materials and equipment). Support activities span the entire organization, as shown in Figure 2. For example, technology development initiatives could attempt to optimize business activities such as fleet management (inbound/outbound logistics), assembly line operation (operations), sales processing (marketing and sales), and help desk (service). In addition, technology optimization may be used in streamlining operations and freeing up resources for the strategic initiatives that drive growth and competitive advantage, and accelerate time to business outcomes (HP, 2007). Margin refers to the potential profit margin that an organization could realize through the sale of its product or service, provided the customer is willing to pay more than the cost of the good sold (i.e. cost of all value chain activities involved, from start to finish, in selling a good).

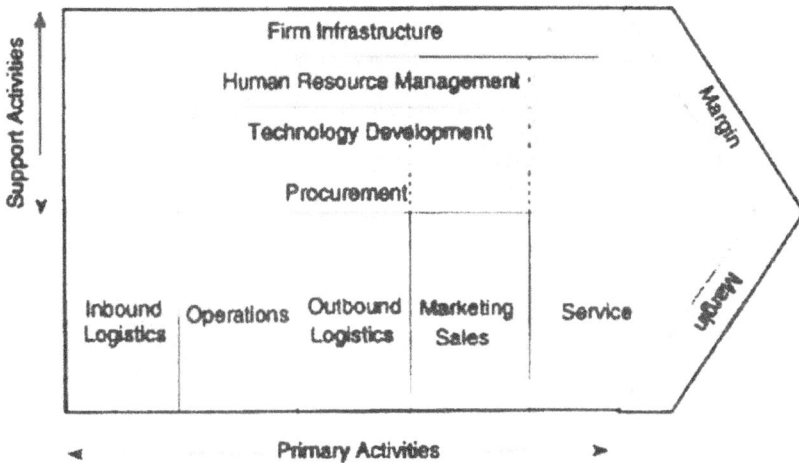

Figure 2: Porter's Value Chain (Source: Porter, 1985)

Organizations search constantly for technological opportunities that could yield a lower cost of the goods sold, increased revenue, or improved customer satisfaction, all of which would translate into strengthening a firm's viability. The next section examines how mobile technology can impact an organization's value chain in these areas.

Mobile Technology and the Value Chain

In a landmark paper, Porter examines the impact of the Internet on the competitive positioning of a firm (Porter, 2001). In this work he argues that the basic tool for understanding the impact of information technologies, such as the Internet, on companies is the value chain. According to this approach, the impact of information technologies on a company can be assessed by examining the influence of such technologies on the primary and support activities in the value chain. Here, we employ this approach to gain an understanding into the impact of mobile technologies on companies.

Primary Activities

Inbound / Outbound Logistics

During these activities a company manages the process of receiving, storing raw materials (i.e. supplies), and distributing finished goods to customers. Supply chain integration and demand chain management are recent extensions in enterprise modeling that require a novel enablement of on-demand information exchanges (Hsu et al., 2007). These information exchanges typically involve a large number of enterprise databases that belong to multiple business partners, and consequently visibility of materials and resources facilitates operational readiness in receiving and delivery timeliness. A current trend highlights the implementation of RFID-augmented systems to integrate enterprise information along the life cycle of a product (Hsu et al., 2007).

RFID tags can be used to track products throughout the entire shipment process (AT&T, 2007c), improving the efficiency of placing new items on the sales floor. For example, after the deployment of their new RFID tagging system, Wal-Mart realized a 19% increase in their use of (RFID tagged) promotional display items. This improvement was attributed to the displays being put up on time and in a correct manner because of the information carried on the RFID tagging system (Hoffman, 2006). In the absence of such visibility, errors can be costly for both inbound logistics, where wrong shipments translate to problems down the supply chain (e.g. meeting outbound deadlines), as well as for outbound logistics where unfeasible order confirmations would otherwise be rejected or rescheduled had real-time inventory data been available at the time the order was being made (Ericson, 2003). Logistics activities can yield strategic business value for a company by lowering distribution costs, reducing inventory, improving customer service, and increasing working capital (Roberts, 2002). Typically neglected, effective inbound logistics can also create value through shorter production and time-to-market cycles of goods produced by the company.

E-Business has been instrumental in generating significant savings during these activities by optimizing processes that previously had been predominantly handled manually. With extensive e-Business applications available in this area, the main driver for using mobile technology is the inherent nature of mobile activities occurring in this segment of a firm's value chain. Receiving raw materials may require the use of a vehicle fleet (e.g. trucks) operated by the company. In this case, wireless fleet management enables real-time visibility of shipment status and performance reporting by providing the location of the shipment's delivery vehicle. For highly valuable products, web-based wireless item-tracking is also possible. Wireless modules are integrated in barcode scanners that allow for automatic registration of shipped products at designated transfer points. This information is then sent wirelessly to a central server for storage. Wireless item-tracking is one of many applications of RFID technology (AT&T, 2007c), making it even more valuable at locations where barcodes cannot be read by fixed devices.

In addition, two-way connectivity between mobile workers (e.g. drivers) and dispatch allows for real-time driving directions, route changes, and delivery schedule updates. General Motors and Siemens are just two of many vendors offering wireless fleet management solutions. These solutions typically make use of the Java2ME platform combined with GPS and GSM/GPRS and other digital networks to enable real-time connectivity between the vehicle, the mobile worker, and dispatch. Solutions are web-based and do not require additional software beyond a web browser (Siemens, 2004).

In addition to the above benefits, fleet management is optimized with integrated wireless solutions. By monitoring a vehicle's status wirelessly, companies are able to improve their situational awareness, security and decision making in tracking and managing shipments… as they move through global supply chains (Biesecker, 2006). This area presents significant opportunity for mobile solution providers, since one-third of U.S. transportation companies have been using mobile technology since 2003 (Collett, 2003). Enabling their entire fleet with wireless tracking and messaging can result in these companies eliminating loading errors, improving productivity, and customer service. In a related example, FedEx adopted handhelds that allowed for data exchange directly with the company's back-end system and its Web-based item-tracking application (Collett, 2006). In another case, Lockheed Martin teamed up with Savi technology to track all of its shipments using RFID-integrated packaging. Not only can customers track the location of their orders, the tags can also be equipped with sensors that measure humidity, temperature, light, and vibration, which let the shippers know the condition of their goods and whether security may have been breached (Biesecker, 2006). 20th Century Fox Home Entertainment International was successful in implementing a mobile strategy that involved the use of wireless devices by sales advisors in the UK market. These mobile professionals switched from the traditional pen-and-paper system to wireless PDAs and Bluetooth-enabled mobile phones for collecting necessary information (e.g. retail store DVD stock levels). An integrated SCM solution that exploited the capability for data synchronization via the wireless Web resulted in improved logistics: the stock replenishment cycle shrunk from three days to one day, while product returns diminished from six-seven weeks to one day. A number of additional benefits apply throughout the value chain, including a five percent increase for on-shelf product availability, a ten percent increase in sales, and a 150-labour-hours-per-month reduction in capturing data (Extended Systems, 2004).

Operations

Operations reflect value-creating activities that transform inputs into final products or services. With emphasis on manufacturing and warehouse activities, mobile technology presents organizations with an opportunity to introduce new or enhanced business processes that would result in greater productivity, efficiency, and effectiveness. It could also result in increased employee satisfaction and lower voluntary turnover (AT&T, 2007c).

The use of mobile technology in manufacturing is particularly evident in the automotive and aerospace industries, where approximately two-thirds of all U.S. based companies are actively using it. For example, General Motors installs wireless computers on forklifts so that drivers can send and receive data, such as work instructions and updates, directly from the factory or warehouse floor. This ability is expected to yield savings in excess of one million dollars at a single GM facility by decreasing use of the forklifts by 400 miles per day, and also in productivity increases as the number of deliveries doubled since implementing the wireless solution (Collett, 2003).

Another opportunity for wireless operations is found in quality control (AT&T, 2007c). MicroElectroMechanical Sensors (MEMS) are being developed that will allow for wireless detection of defects. These sensors will identify out-of-range vibrations in industrial equipment and send, and receive data wirelessly with a range of one thousand feet (Collett, 2003). Their small size, approximately the size of a grain of sand, makes them particularly suited for installation on cumbersome machinery, for which quality inspections would otherwise be lengthy and consequently costly. Predictions of wireless operations in the future show a trend toward machine-to-machine (M2M) communications—or S2S according to Figure 1—for tracking maintenance, service, and status issues (Morley, 2007). By utilizing databases and wireless networking technologies, machines within an operations facility can be monitored automatically, reducing the amount of human labour hours needed to maintain manufacturing equipment.

Real-time wireless asset tracking and inventory visibility is also employed in Operations. Through the use of location-based technologies, e.g. radio frequency identification (RFID) tags and wireless access points (Bryant, 2007), items moved around in a particular facility can be tracked continuously. This allows for faster retrieval of needed items, thus lowering labour costs, increasing productivity and expediting delivery to customers, and subsequently improving customer satisfaction (Collett, 2003). Until recently, the adoption of such technology has been scattered and limited. However decreases in costs and improvement in sensitivity, range and durability have enabled more widespread use of RFID in logistics and operations (Williams, 2004). Powerful players, such as Wal-Mart, have encouraged adoption by requiring their top 100 supplies to place RFID tags on shipping crates and pallets as of January 2005. By the end of 2007, Wal-Mart had over 600 suppliers on board. After two years of RFID implementation, Wal-Mart is starting to reap the benefits, including a 26% reduction in stock-outs along with a plethora of available logistics and sales data (Hoffman, 2006).

BMW is another company that has benefited from RFID implementation by utilizing this technology in its Assembly Finish System to locate any vehicle coming off the assembly line and being parked in any one of 3000 spaces available on site. Similar benefits to those described above are realized through a web-based solution that graphically displays the location of each car on site (WhereNet, 2004).

Inventory visibility is also critical in parts replenishment. Several automakers have implemented wireless solutions that support just-in-time manufacturing processes. Typically, the solution continuously monitors and updates inventory levels as stock is being used, and automatically sends a wireless request specifying the type, volume, and delivery location of a material when needed in real-time. This is an innovative alternative to the traditional paper-based Kanban parts replenishment systems or hardwired electronic call systems, and it offers the twin advantages of low installation costs and unparalleled flexibility in industrial manufacturing environments (WhereNet, 2004). Benefits include lower inventory levels, decreased operating costs, and improved productivity, all of which contributed in a significant Return On Investment (ROI) of less than one year in the case of the Hummer vehicles. Similar benefits were gained by Monroe Truck Equipment by replacing a broad supplier base with a single provider of raw materials (steel), all enabled through the implementation of a novel wirelessly-enabled just-in-time ordering system (Anonymous, 2007).

Moving away from plant operations and manufacturing, mobile technology can offer significant benefits in the service industry as well. YouthPlaces, a nonprofit organization offering youth-related after-school programs, was able to leverage scannable I.D. cards and wireless devices in tracking youth participation in real-time as opposed to experiencing a 30-day lag. This information was then used for activities such as staff scheduling and training (Extended Systems, 2004).

Marketing and Sales

M-Business has been argued by many to be a new channel for commerce. While the objective here is not to support or reject this view, mobile technology certainly enables uniquely two elements of the marketing mix, namely promotion and place (or distribution). Promotion takes the form of wireless advertising and, although it is still at its infancy, it presents significant potential as wireless devices increasingly penetrate the consumer market. Coupled with location-based technology and future built-in sensors and personalization capabilities, wireless promotions can be targeted and more effective. Extending from the promotional opportunities presented, distribution of goods and services to a wireless device is a novel capability, allowing for immediate access/delivery of pertinent data, such as business-related information. By improving the availability of information, mobile workers are more knowledgeable and consequently more productive and effective in satisfying customer needs. Through mobile technology, customer concerns can be addressed immediately by accessing needed resources (e.g. questions on product specifications), without mobile workers having to prepare and carry excessive amount of paper documentation. Finally, in terms of sales, wireless point-of-sale devices enable immediate order fulfillment, reduce the incidence of incomplete transactions (e.g. abandoned shopping carts on the wired Internet), reduce paperwork and waste, improve accuracy of orders, and enhance customer service.

To illustrate these three wirelessly-enabled areas, namely promotion, distribution, and sales, the following examples are cited. Wireless Point-of-Sale (POS) devices are being utilized in retail settings to help employees assist customers on the sales floor without requiring them to wait in long lines for price queries and item availability. On-demand service helps reduce customer turnover, especially during the holidays when large crowds and long lines deter customers (AT&T, 2007a). As an example of promotional activities enabled by mobile technology, SkyGo has been delivering advertisements on wireless devices. Initial consumer feedback has been positive, in particular for time-sensitive coupons from restaurants and media related-promotions such as audio clips for upcoming concerts and movie trailers that further allow users to buy tickets from their wireless Web-enabled phones (News.Com, 2001). While potential benefits of wireless promotions are extensive (e.g. high recall and response rates, reaching clients in a high-growth market sector) (Bergells, 2004), businesses need to place the consumer at the centre of such campaigns and effectively address their concerns. The consumer's ability to personalize the type, volume, and delivery time of advertisements are key success factors in obtaining customer acceptance of this service. In addition to wireless advertising, other forms of wireless promotions include mobile research surveys, e-news sponsorships, and banner ads displayed on wireless Websites.

In terms of distribution, mobile technology provides a new channel for the delivery of simple information such as static web pages, dynamic real-time updates such as location-based traffic information, and rich media such as video streaming of news and movies. Users of web services over mobile phones benefit from anytime connections, enabling activities that required time-sensitive data. Services such as driving directions and weather updates are frequently needed in a mobile setting where a wired connection is not feasible (e.g. while traveling). Mobile e-mail access also helps users increase productivity and respond to important information in a timely manner. These services are being utilized by employers desiring a centrally managed mobile workforce.

For example, MyPrimeTime utilizes wireless distribution of its articles in real-time to members' mobile devices. These life management related articles can be viewed directly on Web-enabled mobile phones or downloaded for future access on a PC via synchronization. To achieve this, MyPrimeTime has partnered with AvantGo to make use of the latter's mobile Internet service (Petersen, 2000).

Drawing from sales applications, Nappi sales force uses wireless devices to send in orders directly to the corporate back-end system, allowing for timely load and schedule updates, which are then automatically forwarded to the plant. Mobile workers are able to save time from placing phone calls to complete an order and the company realizes savings in terms of communication costs. A barcode scanning feature of some mobile devices further reduces the time to complete a sales transaction and eliminates errors as sales people are not required to key in the order (Collett, 2003).

A similar solution implemented by M.R. Williams, a wholesale distributor of various products, involves the use of PDAs for the collection and wireless transmission of critical data (e.g. inventory levels) from retail stores to corporate back-end systems. This integrated approach resulted in sales increases of 34 percent in the first year of the system's use, as well as in freeing up 60 percent of field sales consultants' time by automating product returns and credits. Additional benefits include improved customer satisfaction and inventory control, as well as increased efficiencies and profits (Extended Systems, 2004).

Service

Corporate responsibility does not end with the sale of a product or service. It continues with ongoing support through after-sales activities that aim to maintain or enhance product value. Most often access to information in a timely manner is a critical component in this endeavor. The flexibility of mobile technology is ideal for supporting mobile workers in unplanned situations that call for information with high variability. Equipping mobile workers with knowledge enhances their ability to solve even the most challenging business problems in less time while improving productivity and customer service (IBM, 2004).

The ability to provide time-sensitive information to mobile workers is a growing competitive necessity. Mobile technology can support collaboration through anytime anywhere access to important information including discussions, documents, workflows, notifications, and e-mail, and provides mobile workers with abilities of synchronization, working offline, and flexibility in the device type used. SiteScape addresses this need for information availability through its wireless collaboration solution. Mobile workers have access to corporate information and key business applications such as Customer Relationship Management (CRM), Sales Force Automation (SFA), Supply Chain Management (SCM), and others that improve productivity, reduce cost of communication, and convert captured data into knowledge thus providing a competitive advantage. Similar benefits can be found through push applications such as emails and system updates (e.g. security updates) sent to wireless devices without requiring mobile workers to log in (Ewalt, 2003). As illustrated through the previous examples, most of these benefits can be realized during other activities as well and not only for Service.

Mobile technology can benefit not only businesses and mobile workers, but also customers. Service technicians are equipped with wireless laptops that contain a library of product repair information (e.g. schematics). When a part is required service technicians can immediately place the order wirelessly directly with the supplier (Collett, 2003). This results in a faster repair and consequently improved customer satisfaction. A similar situation is encountered in the health care industry: integrated mobile devices assist health care professionals with checking-up on patients, keeping track of patient status and medications. Mobile integration offers further benefits by helping workers locate necessary equipment and other workers in emergencies, when time is critical (AT&T, 2007a).

In the service industry, caregivers for in-home patient care employed by ST-BNO were equipped with wireless PDAs that provide them with current information and real-time updates in terms of patient schedules and care data. With just one fourth of the work force enabled with the new system, the company has achieved a five percent increase in field service productivity. Additional benefits include fewer errors, shorter billing cycles, lower administrative costs, and an improved level of patient care and satisfaction (Extended Systems, 2004).

Support Activities

Firm Infrastructure

A competitive business environment calls for a firm's ongoing effort to develop competitive advantage. This may be found in any of the following gains:

operational efficiency (e.g. reducing costs, improving communication); innovation (e.g. implementing new business processes); revenue generation (e.g. increased productivity, introduction of new revenue streams); and customer satisfaction (e.g. improved service). Mobility support is a factor that can positively influence any of the above areas. While employee reachability via mobile phones may be a good start, a truly mobile-enabled enterprise emerges only when employees, applications, and infrastructure are fully integrated. A firm's infrastructure supports the entire organization and its value chain through systems and mechanisms for planning and control, such as accounting, legal, and financial services (IDA, 2000). Thus, value creation is optimal when a mobile worker is not only able to receive phone calls, but rather able to communicate with business partners, retrieve data, and analyze it by means of applications made available through a mobile device of any type.

Monitoring and supporting a mobile workforce presents a business challenge that goes beyond traditional management requirements. In a pilot study, AT&T devised a new management strategy for over 5,000 employees, whose mobile communications were carried on a variety of networks with an array of calling plans and pricing schemes. By analyzing the multi-carrier system, it was determined that 23% of mobile employees had calling plans that did not fit their usage. By measuring employee usage against hundreds of calling plans in their Multi-Carrier Solutions platform, AT&T was able to streamline their mobile strategy and reduce average monthly cost for mobile systems by 21% (AT&T, 2007b).

While e-Business technologies were responsible for integrating an organization across its value chain, mobile technology will extend this integration across time and place as well. Two areas that benefit from such wireless platforms are communication and information. Wireless devices enable two-way communication through voice, text messaging (and its variants), e-mail, and video-conferencing. Information availability is supported through the integration of mobile technology on existing Enterprise Resource Planning (ERP) systems and all associated modules, such as accounting (e.g. filing expense claims), manufacturing (e.g. monitoring production levels), and quality (e.g. remote management of information technology) among others. Integration across time and place enables synchronization. Synchronous communication, for example, can be realized more often as the time an employee is not reachable is minimized. Synchronous communication will also translate into faster processing of orders, requests, etc. Finally, integrated systems can increase productivity and subsequent profit. Research in Motion's (RIM) Blackberry provides one such solution for mobile workers requiring access to information and communications. This platform integrates voice, email, SMS, wireless Web, organizer and other productivity applications. The proprietary Enterprise Server seamlessly connects multiple enterprise systems (RIM, 2005).

Another platform offering integrated communications and extensive functionality is IT Solution's m-Power. By utilizing Bluetooth-enabled mobile phones, wireless PDAs and laptops distributed to the company's field service engineers, information technology initiatives including notifications to mobile workers, confirmations of orders, and time sheet management were implemented. This resulted in the following benefits: 50 percent reduction in HelpDesk personnel, 60 percent and 15 percent savings in communication costs to and from field service engineers respectively. Additional benefits include shorter billing cycles and more accurate and reliable expense claim submissions (Extended Systems, 2004).

Human Resource Management

An organization is responsible for employee recruitment, selection, training, development, motivation, and rewards. As employees are an expensive and vital resource to an organization, effective and efficient human resource management (HRM) can add significant value to a firm. Striving for this goal, Motorola decided to redesign its HRM system in an attempt to address present inefficiencies; it was estimated that some employees spent up to 75 percent of their time on administration rather than activities that could be of more value. The solution came in the form of Enet, an HRM system based on Internet technology. This Web-based system, also accessible through wireless devices, allows employees to access critical HR-related information and services anytime anywhere, such as initiating, approving and tracking administrative change requests such as merit increases, leaves of absence and department job changes (Accenture, 2005). Thus, clerical work for HR employees is reduced, subsequently reducing paperwork, and allowing them to concentrate on higher value-adding activities, such as relationship management. Benefits of Enet for employees span the entire organization. For example, mobile workers have a direct line of communication with human resources. As a result, there is improved employee satisfaction and greater credibility for the HRM system given a higher level of consistency than previously achieved through paper-based processes. Savings will be realized in the form of more consistent and efficient processes, cost avoidance, improvements in data integrity and reduced process cycle time, which has dropped from two weeks to two days or less. As a result, the system is expected to pay for itself in just one year. The company also expects Enet to increase employee satisfaction and retention by improving communication and making human resources services more accessible and useful for employees (Accenture, 2005).

Furthermore, in recent years there is a trend towards satisfying the need for a balanced lifestyle or that of increased work-related mobility through telecommuting and flexible work practices. These policies can be achieved by adopting mobile technology. While mobile technology is popular among mobile workers in sales, support and field service, only a few companies have implemented wireless services in HR. However, an organization's workforce is becoming increasingly mobile. For example, the U.S. led the world in 2006 with 68% of its workforce being mobile and it is estimated to reach 75% by 2011 (IDC, 2008). At a global level, the mobile workforce is expected to grow by more than 20 percent, with 878 million people working remotely by 2009 (Gosling, 2007) and 1 billion doing so by 2011 (IDC, 2008).

Consequently, wireless HR solutions will become a critical component in successful HRM strategies (Roberts 2001). For example, Wireless-i offers complete solutions for expense and time sheet management that allow employees to enter work-related claims and up-to-date time sheet information easily anytime anywhere (Wireless-I, 2005). By monitoring time utilization and expenses, these solutions allow organizations to reduce HR-related costs, empower employees, improve employee satisfaction, and improve productivity (AT&T, 2007d).

Similar control over field service representatives (FSRs) was desired by Valspar, a leader in the paint and coatings industry. By using their wireless PDAs to scan retail store inventory and update back-end systems, FSRs were tracked in terms of their location and time spent for each job. This feature resulted in better time management by FSRs and in a decrease from three weeks to two days for generating results on ad-hoc requests (Extended Systems, 2004).

Technology Development

Activities focusing on technology development add value to an organization by introducing innovative technology that improves services, products, and business processes. Hence, technology development is an important catalyst for competitive advantage. The latest trend in technology development involves m-Business, where the utilization of mobile technology can potentially reap the above benefits thereby strengthening a firm's value chain. Whether in-house or outsourced, development of wireless solutions can target any of the primary activities and/or their linkages. At the same time, mobile technology can enhance the research process with real-time access to pertinent information regardless of time and/or geographic location, such as real-time consumer feedback transmitted from the user's device (e.g. wireless survey) and wireless access to the organization's knowledge base and knowledge directory. Communication may be initiated by the user or it may be set up to occur automatically between a mobile device and the network at specified times. In addition, mobile technology can foster product development by providing a flexible yet powerful platform for collaboration across locations. Furthermore, the use of the Internet has been shown to have a significant positive impact on Research and Development (R&D) (Linder and Banerjee, 2005). Since m-Business delivers the Internet wirelessly, the benefits gained from e-Business are transferable, thus creating additional value for the organization (caution is needed given the novel usability issues associated with mobile technology). Expected benefits of mobile technology, both current and emerging as in the case of WiMax (AT&T, 2006). in Technology Development include improved productivity through greater accuracy (as calibration can be constantly corrected), improved production times due to reduced downtime, greater flexibility in production times and volumes (DTI, 2005).

One company that has utilized mobile technology in this context is 3Com. The company was able to leverage these wireless networks to strengthen the relationships among team members by improving the communication amongst them and the availability of information to them (3Com, 2005).

Improved communication may also be realized through On Demand Mobile Conferencing (ODMC), a solution offered by Zeosoft, a provider for mobile infrastructure software and application development technologies. ODMC enables real-time exchange of information through text messages, file sharing, and live group discussions with white boarding capabilities on a virtual work space accessed by wireless devices. The solution improves existing business processes, increases employee productivity, and reduces the cost of conducting meetings (ZeoSoft, 2005).

The ability to gain access to Personal Information Management (PIM) (e.g. e-mail, contact lists) and groupware data was also enabled by First Command's wireless solution. First Command, an international financial management company, implemented a system that allowed for real-time synchronization of sales associates' mobile devices, which not only enabled anytime, anywhere collaboration via the corporate Microsoft Exchange system, but also resulted in savings for each associate of up to three hours per day (Extended Systems, 2004).

Procurement

This support activity encompasses all purchasing transactions for goods and services. Optimal conditions include the lowest price and highest quality for what is being purchased. Mobile technology can add value by enhancing current electronic procurement practices, such as web-based order fulfillment. Transactional cost savings, increased flexibility, and customer satisfaction are a few of the expected benefits realized when enabling an organization with wireless procurement. Corrigo, a service management solutions provider, offers an application to property managers that enables field technicians to order repair parts through a WAP-enabled mobile device. Eliminating the burden of searching through catalogs for part numbers, followed by phone calls to place an order, apartment maintenance and repair workers can directly access supplier data and order needed parts. In addition, this IT solution brings property managers closer to customers (i.e. residents) by allowing them to enter a service call either by phone or online instead of having them visit the property management's office, and relaying that information immediately to the mobile repair worker. The application also allows residents to track the work order status, while property managers are given visibility to maintenance personnel activities (Moozakis, 2000).

Elcom International, on the other hand, has extended their Internet Procurement Manager to wireless devices. Initially capable only for routing and approvals, eMobileLink enables e-mail notifications of requests for quotes (RFQs), downloading and viewing RFQs, and approving/rejecting them from a wireless device, while integrating settlement capabilities (Ferguson 2001).

A new trend in mobile workforce management is Fixed-Mobile Convergence (FMC), which utilizes a centralized management structure which oversees mobile employees. By using equipment that operates over a variety of networks including cellular, Wi-Fi, and possibly WiMAX in the near future, employees can access and transmit data. When associated with an office private branch exchange (PBX), FMC-enabled devices offer all the functionality of an office phone and laptop computer while allowing the freedom of wireless networks and cellular coverage (Winther, 2007).

In terms of order entry, Zync Solutions, a provider of web-hosted software solutions, equipped field representatives with mobile devices for scanning bar codes instead of placing orders manually, as well as recording any additional information that may be obtained during the store visit. With time savings of 30 percent (i.e. 150 labour hours per month), the information is sent up the value chain wirelessly via the corporate back-end system. Additional benefits include improved information flow, efficiency, productivity, reporting accuracy, response time to retailers' needs, which subsequently improve sales and a faster return on investment (ROI) (Extended Systems, 2004).

Discussion

The foregoing discussion found in the previous section was summarized in Figure 3. Figure 3 describes an organization's primary and support activities in terms of both representative applications currently found in industry, as well as the interaction types (included between brackets) that convey which interactions, from those depicted in Figure 1, are being enhanced by the listed application. Cognizant of these value-adding mobile technologies, managers can then better leverage m-Business to support and enhance both the primary and support activities of an organization's value chain contributing to a firm's overall competitiveness.

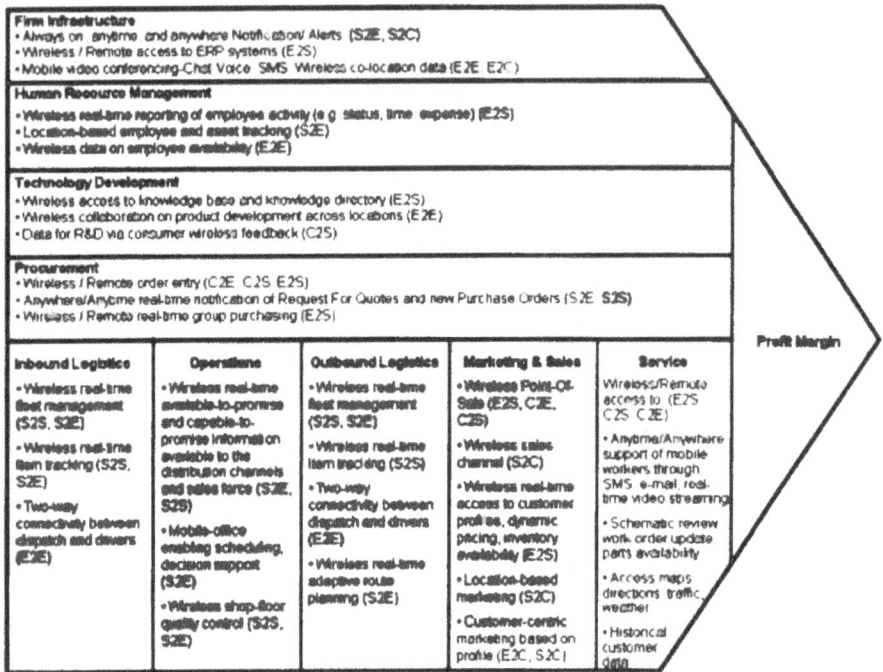

Firm Infrastructure
- Always on, anytime and anywhere Notification/ Alerts (S2E, S2C)
- Wireless / Remote access to ERP systems (E2S)
- Mobile video conferencing-Chat Voice SMS Wireless co-location data (E2E E2C)

Human Resource Management
- Wireless real-time reporting of employee activity (e.g. status, time expense) (E2S)
- Location-based employee and asset tracking (S2E)
- Wireless data on employee availability (E2E)

Technology Development
- Wireless access to knowledge base and knowledge directory (E2S)
- Wireless collaboration on product development across locations (E2E)
- Data for R&D via consumer wireless feedback (C2S)

Procurement
- Wireless / Remote order entry (C2E C2S E2S)
- Anywhere/Anytime real-time notification of Request For Quotes and new Purchase Orders (S2E S2S)
- Wireless / Remote real-time group purchasing (E2S)

Inbound Logistics	Operations	Outbound Logistics	Marketing & Sales	Service
• Wireless real-time fleet management (S2S, S2E)	• Wireless real-time available-to-promise and capable-to-promise information available to the distribution channels and sales force (S2E, S2S)	• Wireless real-time fleet management (S2S, S2E)	• Wireless Point-Of-Sale (E2S, C2E, C2S)	Wireless/Remote access to (E2S C2S C2E)
• Wireless real-time item tracking (S2S, S2E)	• Mobile-office enabling scheduling, decision support (S2E)	• Wireless real-time item tracking (S2S)	• Wireless sales channel (S2C)	• Anytime/Anywhere support of mobile workers through SMS e-mail, real-time video streaming
• Two-way connectivity between dispatch and drivers (E2E)	• Wireless shop-floor quality control (S2S, S2E)	• Two-way connectivity between dispatch and drivers (E2E)	• Wireless real-time access to customer profile, dynamic pricing, inventory availability (E2S)	• Schematic review work order update parts availability
		• Wireless real-time adaptive route planning (S2E)	• Location-based marketing (S2C)	• Access maps directions traffic weather
			• Customer-centric marketing based on profile (E2C, S2C)	• Historical customer data

Profit Margin

Figure 3: Prominent m-Business applications in the Value Chain

By exploring the impact of mobile technologies on the various components of the value chain and through citing extensive industry examples, this paper has demonstrated the potential of such technologies. The applications outlined in the previous section can be generalized and grouped according to the following classification:

Asset Tracking

Referring to either physical objects (e.g. merchandise) or human resources (e.g. employees), these applications allow organizations to access tracking information. The organization then leverages the assets' visibility for optimizing processes (e.g. timeliness of deliveries). The function of tracking employees could also be combined with the capability for continuous communication thereby increasing the value of these applications (e.g. a mobile worker equipped with a GPS enabled mobile phone).

Data Access

Access to time-sensitive information could enhance an organization's efficiency and effectiveness resulting in competitive advantages. Information could either be pushed to the employees, business partners, and/or consumers (e.g. through SMS), or pulled by employees from remote locations (e.g. field technicians requiring specifications for various jobs). Data access may also optimize an organization's data management, with collaboration applications that support knowledge sharing and increase knowledge flow.

Automation

Mobile technology can be used to automate some tasks previously performed by employees. Benefits for an organization may include lower workforce requirements, improved employee time allocation, and improved quality by automating processes and reducing employee errors. One company that implemented such mobile workforce automation processes is Intermountain Gas Company (IGC). IGC serves more than 275,000 natural gas customers across Southern Idaho and employs 350 people in seven district offices (IGC, 2008; Itron, 2005). Unlike many utilities, IGC did not have an integrated dispatch system for work orders rather routed orders either by paper or radio. By enabling dispatch and field service workers to communicate and share data in real time through a wireless, web-based mobile communications and automated solution, IGC improved emergency response by quickly identifying the nearest field representative with appropriate skills; decreased fleet mileage through tighter, more efficient routing and streamlined order processing by eliminating reams of paper orders and reducing data entry errors. Employee productivity and customer satisfaction increased while the costly paperwork and time associated with traditional manual work order processing was eliminated (Itron, 2005).

Despite the above applications and associated benefits of mobile technology, it is still in its infancy and companies are faced with the dilemma of why, and if so, when they should invest in it. The decision will depend on many factors, one of which is whether the organization's workforce needs to be or is already mobile. In this case, opportunities arise according to the environment in which it operates. Within a B2B and B2E environment, the value propositions are similar in that mobile technology and the corresponding applications aim to improve the productivity of the parties involved, while the focus varies between the two settings. In B2B it is the efficiency and effectiveness of the interactions between organizations that is of interest. In B2E the efficiency and effectiveness of a single worker and/or a team can be enhanced by wireless solutions that help increase productivity, streamline administrative processes, and build competitive advantage by simplifying and improving the effectiveness of collaboration.

In addition to the above considerations, there are several concerns that arise with using mobile technology in a business setting. Such concerns exist at the level of employees, organizations, and even society at large, and include the following

Employees

Poor ease of use and low perceived usefulness may be deterrents in workers' adoption of a newly implemented technology (Davis et al., 1989). This usability concern is related to the concern for compatibility between current mobile technology capabilities and employee expectations, values and experiences. Also, privacy concerns may arise, as the content of an employee's communication, but also his/her whereabouts may be perceived as intrusive and as a threat to their individual privacy.

Organization

Given the unique nature of mobile technology and its vulnerability (Coursaris et al., 2003), concern regarding the safety of information exchanged over a wireless network increases with the degree of interaction and the sensitivity of the information exchanged (Rogers, 1995). In addition to security, there is concern over the reliability of the technology. Connection quality should be maintained for the specified network coverage. The inherent concern here is that loss of the connection can result in loss of data (Nielsen, 2000). Lastly, organizations are faced with the concern over the cost of implementing mobile technology and the expected return on that investment. As some benefits may be intangible and difficult to quantify (e.g. improved communication, timely decision making, improved customer satisfaction through increased responsiveness, etc.), it may be challenging for a business to have sufficient evidence in support of adopting mobile technology.

Society

Social skepticism around the growing use of mobile technology is in part due to the confusion over its effects on people's health. Studies have been inconclusive on whether this technology can be potentially harmful in the long-term, but in the absence of a clear answer there is apprehension towards use of such systems. Concerns over an individual's privacy and security may also deter them from using mobile solutions put forth by businesses. For example, location tracking may be perceived as threatening, both in the context of unsolicited messages/advertising and physical safety, as this information could be dangerous if intercepted. Furthermore, anytime and anywhere access offered through mobile technology provides employees with valued flexibility, but further blurs the line between work and home. Since mobility may provide 24/7 access to employees, expectations of 24/7 availability and responsiveness may also surface. This may have detrimental effects on the quality (and quantity) of leisure time and home life.

Conclusion

This paper examines the potential for mobile technologies to provide value to various business activities. Resulting benefits of mobile technology implementations may include improved productivity through enhanced process efficiency and effectiveness, as well as improved customer service. Organizations, however, need to be cognizant of potential concerns among employees and society at large in their assessment and implementation of such technologies. This paper can serve managers as a go-to resource during their initial consideration of mobile technology. Rather than making multiple choices and adopting various technological standards, a comprehensive consideration of the organization's value chain can provide a holistic representation of the company's needs. Having identified such an aggregate set of needs, information technology (I.T.) managers can proceed with the adoption of integrated systems that either include or entirely consist of mobile technologies and span multiple areas of the organization supporting a subset of services presented in this paper and delivering maximum value. In addition, I.T. managers were provided in this paper with a different lens that could be used during their assessment of I.T. resource needs. Moving beyond the individual user's needs and considering each individual's potential interactions with other employees and systems, an enhanced set of user requirements is produced from which, again, I.T. choices pertaining to mobile technology adoption are likely to become less risky and may have a higher return on investment.

While the value-chain framework provided in this paper can help managers employ new mobile technologies or assess the value and appropriateness of existing mobile applications, there are several fruitful areas for further investigation. Future research could delve more deeply into each primary or support activity to explore specific advantages and obstacles across various industries. Detailed case studies can be examined to provide managers with concrete best practices in their industry or comparable industries. Additionally, the value chain framework could be used as a lens to understand the current mobile technology platforms and how they support specific needs and expectations of employees, trading partners and customers. Lastly, empirical data could be gathered to provide further evidence of mobile technology usefulness for various value chain activities. Such data could focus on the perceived value of such technologies from various stakeholder perspectives.

Acknowledgements

The authors thank the journal's reviewers for their comments. An earlier version of this paper was presented in the 2006 International Conference on Mobile Business (ICMB), Copenhagen, Denmark, June 26-27, 2006.

References

3Com (2005). 3Com Goes Wireless To Improve Employee Productivity. Retrieved February 5, 2008, from http://www.3com.com/solutions/en_US/casestudy. jsp?caseid=137286

Accenture (2005). Internet-Based Human Resource Solution Empowers Motorola Employees. Retrieved on January 28, 2005, from http://www.accenture. com/xd/xd.asp?it=enweb&xd=industries%5Ccommunications%5Chightech%5Ccase%5Chigh_motorola.xml

Anonymous (2007). Just-in-time model eliminates Posat Monroe Truck Equipment. Purchasing, 136, 11, 32–37.

AT&T. (2007a). Vertical Applications for Wireless LANs. AT&T Knowledge Ventures. Retrieved October 25, 2007, fromhttp://www.business.att.com/resource.jsp?repoid=ProductCategory&repoitem=e b_mobility&r type=Whitepaper&rvalue=vertical_applications_for_wireless_lans&download=yes&segment=who le

AT&T. (2007b). Making the Case for Enterprise Mobility. AT&T Knowledge Ventures. Retrieved October 25, 2007, from http://www.business.att.com/content/whitepaper/POV-TCO-mob_11495_V03_11-07.pdf

AT&T (2007c). Radio Frequency Identification. AT&T Knowledge Ventures. Retrieved October 25, 2007, from http://www.business.att.com/content/whitepaper/radioidentification.pdf

AT&T (2007d). Market brief: The Age of the Wireless LAN. AT&T Knowledge Ventures. Retrieved October 25, 2007, from http://www.att.com/Common/merger/files/pdf/wireless_LAN.pdf

AT&T (2006). Laying the Groundwork for WiMAX. AT&T Knowledge Ventures. Retrieved October 25, 2007, from http://www.business.att.com/content/article/wimax_gndwk_pov.pdf

Bergells, L. (2004). Wireless Advertising Plan in Six Steps. Maniactive. Retrieved October 25, 2007, from http://www.maniactive.com/wireless.htm.

Biesecker, C. (2006). Lockheed Martin Forms Savi Group to Zero In On Supply Chain Solutions. C41 News. November 23, 2006. 1. Retrieved November 15, 2007.

Bryant, J. (2007). WiFi on the plant floor. Control Engineering. 54, 8, 8.

Collett, S. (2003). Wireless Gets Down to Business. Computer World, May 23, 2003. Retrieved on January 28, 2005, from http://www.computerworld.com/mobiletopics/mobile/story/0,10801,80864,00.html

Coursaris, C., and Hassanein H. (2002). Understanding m-Commerce: A Consumer-Centric Model. Quarterly Journal of Electronic Commerce, 3, 3, 247–272.

Coursaris, C., Hassanein, K., & Head, M. (2003). M-Commerce in Canada: An Interaction Framework for Wireless Privacy. Canadian Journal of Administrative Sciences, 20, 1, 54–73.

Davis, F. D., R. P. Bagozzi, and P. R. Warshaw (1989). User acceptance of computer technology: A comparison of two theoretical models. Management Science, 35, 8, 982–1003.

DTI (2005). Wireless & Mobile. Department of Trade and Industry. Retrieved on October 25 from http://www.dti.gov.uk/bestpractice/technology/wireless-mobile.htm

Ericson, J. (2003). Considering Inbound Logistics. Line 56, August 23, 2003. Retrieved on July 30, 2004, at http://www.line56.com/articles/default.asp?ArticleID=4936&ml=2.

Ewalt, D. (2003). OneBridge Mobile Groupware lets companies support remote workers across multiple networks, hardware, and platforms. InformationWeek, August 4, 2003. Retrieved on January 28, 2005, from http://www.informationweek.com/story/showArticle.jhtml?articleID=12808297

Extended Systems (2004). Success Stories. Retrieved on January 28, 2005, from http://resolution.extendedsystems.com/esi/products/mobile+data+management+products/shared/su ccess+stories/_success+stories.htm.

Ferguson, B. (2001). E-Procurement Goes Wireless. eWeek, February 7, 2001. Retrieved on January 28, 2005, from http://www.zdnet.com.au/news/communications/0,2000061791,20156914,00.htm.

Gosling, A. (2007). Lumension Integration For Enterprise Security. Mobilise, December 14, 2007. Retrieved on March 4, 2008, from http://www.mobilised.com.au/content/view/1205/96/.

Hoffman, W. (2006). Wal-Mart Tags Up. Traffic World. August 16, 2006, 1.

Hsu, C., Levermore, D., Carothers, C., Babin, G. (2007). Enterprise Collaboration: on-demand information exchange using enterprise databases, wireless sensor networks, and RFID systems. IEEE Transactions on Systems, Man, and Cybernetics—Part A: Systems and Humans, 37, 4, 519–532.

HP (2007). HP BTO software: Accelerate time to business outcomes—White Paper. Revised June 1, 2007. Retrieved on March 3, 2008, from https://h10078.www1.hp.com/cda/hpms/display/main/   hpms_content.jsp?zn=bto&cp=1-11-%5E4864_4000_100__

IBM (2004). RFID solution for supply chain management and in-store operations. Retrieved from http://www-1.ibm.com/industries/wireless/doc/content/solution/1025230104.html

Introducing a truly Mobile Server (2005). Retrieved on January 28, 2005, from http://www.zeosoft.com/htmlsite/downloads/Product_Overview.pdf.

IDA (2000). Infocomm21: Leadership Dialogue. IDA Singapore 2000, August 2000. Retrieved on January 28, 2005, from http://unpan1.un.org/intradoc/groups/public/documents/APCITY/UNPAN011538.pdf.

IDC (2008). IDC Predicts the Number of Worldwide Mobile Workers to Reach 1 Billion by 2011. Press Release, January 15, 2008. Retrieved on March 3, 2008, from http://www.idc.com/getdoc.jsp?pid=23571113&containerId=prUS21037208.

IGC (2005). About IGC. Retrieved on March 3, 2008, from http://www.intgas.com/aboutigc/aboutigc.html.

Itron (2008). Itron to Provide Service-Link® Mobile Workforce Automation Technology to Intermountain Gas Company. Press Release, June 27, 2005. Retrieved on March 3, 2008, from http://www.itron.com/pages/news_press_individual.asp?id=itr_000131.xml.

Linder, J. and Banerjee, P. (2005). Research and Development in the 21st Century: Web-Enabled Innovation Comes of Age. Accenture. Retrieved on October 25, 2007, from http://www.accenture.com/xd/xd.asp?it=enweb&xd=ideas%5Coutlook%5Cpov%5Cpov_randd.xm l.

Moozakis, C. (2000). Procurement App Will Go Wireless. InternetWeek, December 11, 2000. Retrieved on January 28, 2005, from http://www.internetweek.com/ebizapps/ebiz121100-2.htm.

Morley, D. (2007). M2M—The New Robotics. Manufacturing Engineering, 138, 4, 144.

News.com (2001). Cell phones: The next great ad conduits? Retrieved from http://news.com.com/Cell+phones+The+next+great+ad+conduits/2009-1033_3-254140.html

Nielsen, J. (2000). Designing Web Usability: The Practice of Simplicity. Indianapolis, Indiana, New Riders Publishing.

Petersen, L. (2000). MyPrimeTime, Inc. Gains Wireless Distribution Through AvantGo. MyPrimeTime, November 2000. Retrieved on January 28, 2005, from http://www.myprimetime.com/misc/press/Avant_Go.shtml.

Porter, M. (1985). The value chain and competitive advantage, Chapter 2 in Competitive Advantage: Creating and Sustaining Superior Performance. Free Press, New York, 33–61.

Porter, M. (2001). Strategy and the Internet. Harvard Business Review, 79, 3, March 2001, 21 pgs.

RIM (2005). Products. Retrieved on October 25, 2007, from http://www.rim.com/products/index.shtml

Roberts, B. (2001). HR Unplugged: Wireless technology could help practitioners better serve an increasingly mobile work force—HR Technology: Systems & Solutions—Human Resources—Statistical Data Included. HR Magazine, December 2001. Retrieved on January 28, 2005, from http://www.findarticles.com/p/articles/mi_m3495/is_12_46/ai_81393642/pg_2.

Roberts, M. (2002). Inbound Logistics Recognizes Schneider Logistics as Top Logistics IT Provider. Schneider Logistics, May 20, 2002. Retrieved on January 28, 2005, at http://www.schneiderlogistics.com/company_info/news_releases/ibltop100_0502.html.

Rogers, Everett M. (1995). Diffusion of Innovations. New York: The Free Press.

Siemens (2004). GSM/GPRS Modules keep a Constant Watch on Goods and Products. Retrieved on October 25, 2007, from http://communications.siemens.com/cds/frontdoor/0,2241,hq_en_0_2245_rArNrNrNrN,00.html.

Turban, E., Lee, J., Warketin, M., and Chung, M. (2002). Electronic Commerce: A Managerial Perspective, Prentice Hall, page 867.

WhereNet (2004). Hummer Factory Implements WhereNet's Wireless Solutions to Enhance Assembly

Line Operations and Expedite Production. WhereNet, March 9, 2004. Retrieved on January 28,

2005, from http://www.wherenet.com/pressreleases/pr_03_09_2004.html.

White, A. and, Breu, K. (2005). Mobile Technologies in the Supply Chain: Emerging Empirical Evidence of Applications and Benefits. Proceedings of the 26th World Congress on the Management of E-Business, Hamilton, Ontario, Canada, January 19–21, 2005.

Williams, D. (2004). The Strategic Implictions of Wal-Mart's RFID Mandate. Retrieved from http://www.directionsmag.com/article.php?article_id=629&trv=1 &PHPSESSID=8beb74b1215e23 26d82ac11e775091c5

Winther, M. (2007). Fixed-Mobile Convergence: Lowering Costs and Complexity of Business Communications.

Wireless-i (2005). Products. Retrieved from http://www.wireless-i.com/ourProducts. asp.

ZeoSoft (2005). A New Approach To Enterprise Wireless Strategies. Retrieved on October 25, 2007, from http://www.zeosoft.com/htmlsite/downloads/Product_ Overview.pdf.

Cost Effectiveness of Community-Based Therapeutic Care for Children with Severe Acute Malnutrition in Zambia: Decision Tree Model

Max O. Bachmann

ABSTRACT

Background

Children aged under five years with severe acute malnutrition (SAM) in Africa and Asia have high mortality rates without effective treatment. Primary care-based treatment of SAM can have good outcomes but its cost effectiveness is largely unknown.

Method

This study estimated the cost effectiveness of community-based therapeutic care (CTC) for children with severe acute malnutrition in government primary health care centres in Lusaka, Zambia, compared to no care. A decision tree model compared the costs (in year 2008 international dollars) and outcomes of CTC to a hypothetical 'do-nothing' alternative. The primary outcomes were mortality within one year, and disability adjusted life years (DALYs) after surviving one year. Outcomes and health service costs of CTC were obtained from the CTC programme, local health services and World Health Organization (WHO) estimates of unit costs. Outcomes of doing nothing were estimated from published African cohort studies. Probabilistic and deterministic sensitivity analyses were done.

Results

The mean cost of CTC per child was $203 (95% confidence interval (CI) $139–$274), of which ready to use therapeutic food (RUTF) cost 36%, health centre visits cost 13%, hospital admissions cost 17% and technical support while establishing the programme cost 34%. Expected death rates within one year of presentation were 9.2% with CTC and 20.8% with no treatment (risk difference 11.5% (95% CI 0.4–23.0%). CTC cost $1760 (95% CI $592–$10142) per life saved and $ 53 (95% CI $18–$306) per DALY gained. CTC was at least 80% likely to be cost effective if society was willing to pay at least $88 per DALY gained. Analyses were most sensitive to assumptions about mortality rates with no treatment, weeks of CTC per child and costs of purchasing RUTF.

Conclusion

CTC is relatively cost effective compared to other priority health care interventions in developing countries, for a wide range of assumptions.

Background

Children aged under five years with severe acute malnutrition (SAM) in Africa have high mortality rates without effective treatment [1-5]. Hospital inpatient treatment of SAM can reduce mortality [5], but in developing countries hospital treatment is too inaccessible and costly for most children with SAM. Community-based therapeutic care (CTC) is a recent model for early diagnosis and treatment of SAM in ambulatory primary health care settings. The key nutritional component of SAM treatment is ready to use therapeutic food (RUTF). This is a nutrient-dense food with a nutrient content/100 kcal that is similar to F100 milk, the diet recommended by the World Health Organization (WHO) in the recovery phase of the SAM treatment [6]. A major advantage of RUTF over F100 is that it contains little water and is thus resistant to microbial contamination, and suitable for storage and use at home without refrigeration [7]. Other key components of CTC are simplified clinical protocols, decentralised provision, community mobilisation and high population coverage [8]. CTC also includes supplementary feeding for moderate malnutrition, which is not considered in this study [8]

Several home-based RUTF programmes in developing countries have shown good outcomes [8-10]. However there have been few controlled trials comparing mortality rates with other treatments [11,12]. No trials have prospectively compared CTC with no treatment, which would be unethical. Although resources constraints are critical for the expansion of CTC, we are aware of only one published study reporting original data on costs of ambulatory treatment of severe acute malnutrition in a developing country [13]. That trial, with 437 children in Bangladesh in 1990 and 1991, showed that inpatient care cost $156 per child, day care cost $59 and domiciliary care cost $29.

Lusaka, Zambia, provides an innovative example of large scale provision of CTC through government primary health care centres. Since 2005 the Lusaka District Health Management Team (LDHMT), which is responsible for the city's 25 primary health care centres, has steadily expanded CTC provision by its staff working in these health centres. By January 2008, 21 of the 25 LDHMT health centres were providing CTC. CTC clinics were set up within each health centre, each staffed by a nurse, a health educator and two volunteers. Nurses were trained to diagnose SAM in children under 5 years of age, by measuring mid upper arm circumference (MUAC) and examining children for bilateral pedal pitting oedema. SAM was defined as MUAC of 11 cm or less, or bilateral pitting oedema [8]. Children were treated with RUTF of 200 kcal/Kg/day, broad-spectrum antibiotics, vitamin A, folic acid, anti-helminthics and, if indicated, anti-malarial treatment [8]. They were then asked to return weekly until they had recovered. Recovery was defined as having MUAC > 11 cm, weight gain and no oedema for at least two weeks, and clinically well [8]. Children with initial MUAC of 11 cm or less at admission were supposed to receive at least 8 weeks treatment, although in practice duration of treatment varied. Children were referred to hospital for inpatient care if they failed to respond to treatment, deteriorated or were severely ill and required hospital care.

Twenty volunteers attached to each health centre screened children at the health centres and in the community and referred those with SAM to CTC. Government and private sector nurses and traditional health practitioners working near to each health centre were trained and encouraged to identify and refer SAM cases. Popular art theatre discussions were held to raise community awareness of

SAM and CTC. Valid International, a company specialising in nutritional research in developing countries, helped initiate the programme and provided technical support to the LDHMT for implementation and staff training. RUTF was manufactured in Lusaka or imported, and delivered to LDHMT medical stores free of charge.

The aims of this study were 1) to describe the outcomes of CTC in Lusaka, 2) to estimate the costs of CTC and 3) to estimate the effectiveness and cost effectiveness of this type of CTC, compared to no treatment. The reasons for comparing CTC to no treatment were, first, to enable the cost-effectiveness of CTC to be compared to any other health care intervention and, second, because comparable data on costs of alternative ways of treating SAM in this population were not available.

Methods

The study was a cost effectiveness analysis based on a decision tree model [14]. Cost and cost effectiveness were considered from the perspective of health services. The existing model of care was compared to a hypothetical alternative of providing no treatment [15]. Household and societal costs of illness and care were beyond the scope of this study.

Decision Tree

The structure of the decision tree is shown in Figure 1. The square represents choice, circles represent chance (probabilities) and triangles represent outcomes. There are two options: "do nothing" or "CTC." For each option, various things could happen to each child, leading ultimately to recovery or death. For the "do nothing" option, death rates differed according to whether children were HIV infected or not. For the "CTC" option, HIV status was not considered, because the effects of HIV/AIDS were already incorporated into known CTC outcomes, and because the HIV status of most children receiving CTC was not known. Children receiving CTC in health centres could have one of the 4 outcomes known to the CTC programme. Children referred to hospital, and children who defaulted, then either died or recovered. For each option, the probability of each outcome was entered into the model to calculate expected rates of death or recovery. For the CTC option, costs of CTC and of hospital treatment were also entered into the model. The "do nothing" option was assumed to cost health services nothing. Effectiveness of CTC was calculated as the difference in death rates between the two options. The cost of CTC, divided by the effectiveness of CTC, is the incremental cost effectiveness ratio, expressed in dollars per life saved. Assuming that each child who recovers has a life expectancy of 33.3 disability adjusted life years [5], cost effectiveness was also expressed in dollars per DALY gained.

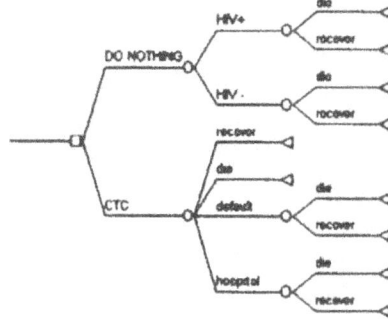

Figure 1. Decision tree.

Model Parameters: Probabilities of Outcomes

Model parameters are shown in Table 1. The primary outcomes of interest were mortality up to year after developing SAM, and expected DALYs after surviving one year.

Table 1. Assumptions, distributions and sources of model parameters

Parameter	Mean	Standard error[a]	Source and comments
Outcomes			
Do nothing option			
Mortality without CTC (HIV-)	0.18	0.045	[1-4] SE assumed.
Relative risk of death with HIV, no CTC	2.0	0.5	[18] SE assumed
Prevalence of HIV in under fives	0.15	0.0375	[16,17] SE assumed.
CTC option			
Death rate during CTC	0.026	0.0032	Programme data.
Proportion defaulting from CTC	0.172	0.0075	Programme data.
Death rate in defaulters from CTC	0.058	0.029	Assumed. SE set so 95% CI is +/- 100% of mean
Hospital referral rate from CTC	0.059	0.0047	Programme data.
Death rate in hospital	0.37	0.093	UTH data. SE assumed
Mortality within a year of recovery	0.0364	0.0091	[19]
Expected DALYs if child recovers	33.3	NA	[5]
Costs (CTC option only)			
No. weeks of CTC - recovered	6.6	1.6	Programme data.
No. weeks of CTC - referred	4.8	1.1	Programme data.
No. weeks of CTC - died	3.6	1.6	Programme data.
No. weeks of CTC - defaulted	5.1	1.5	Programme data.
Cost per health centre visit	$4.24	$1.06	LDMHT. SE assumed.
Cost per kg of RUTF	$6.10	$1.53	Valid International. SE assumed.
Kg of RUTF per week per child	1.90	0.016	Programme data.
Cost of community mobilisation per child	$1.06	$0.27	LDMHT. SE assumed
Valid cost per child	$68.69	$17.17	Valid International. SE assumed.
Cost per day in hospital	$41.35	$10.34	[24] SE assumed.
Days in hospital	14	3.5	UTH data. SE assumed.

[a] SE standard errors for probabilistic sensitivity analysis; normal distributions assumed.
CI confidence interval. $ international dollars, year 2008. NA not applicable. UTH Lusaka University Teaching Hospital

For the 'do nothing' option, expected mortality was based on evidence from a review of child mortality rates associated with malnutrition in developing countries [1]. In particular we based mortality rates on rigorous community based cohort studies conducted in Malawi [1,2] and Uganda [3,4] in the late 1980s that used MUAC as a predictor of mortality. In the Lusaka CTC population, the median MUAC in children with MUACs of 11 cm or less was 10.6 (interquartile range 10.0–10.8) cm and in children with oedema the median was 12.0 (interquartile range 11.2–13.0) cm. Mortality rates with bilateral pedal oedema were assumed to be the same as with MUAC of 11 cm or less, as we found in this CTC programme. At the time of the cohort studies [1-4], for such children minimal treatment was available and the prevalence of HIV was negligible. Non-CTC children were therefore stratified by HIV status, to account for the increased death rate with HIV, the prevalence of which has increased over the past 20 years. The HIV prevalence estimates was based on numbers of infected children in Zambia, from UNAIDS [16] and numbers of children aged under 5 from the Zambia census [17]. Mortality with HIV was assumed to be double that without HIV [18].

For the CTC option, outcomes at the end of health centre care were death, recovery, referral to hospital, or default (Table 1). These outcomes were known from programme data for 2523 patients treated from September 2005 to September 2007. Among children referred from CTC to hospital, the death rate was assumed to be 37%, which was the death rate in the University Teaching Hospital (UHT) acute malnutrition ward (personal communication Dr B Amadi, UHT paediatrician). The death rate among children who defaulted from CTC was assumed to be the same as for all other children, including those referred to hospital, because they had similar prognostic characteristics. That is, among children who defaulted, mean initial MUAC was 11.2 cm and 25.2% had oedema; among children who did not default, mean initial MUAC was 11.4 cm and 25.3% had oedema. The cohort estimates of mortality used for the "do nothing" option were based on one year of follow up, but CTC programme data were based on an average of 7 weeks of follow-up. To be able to compare annual mortality rates between the two options we therefore assumed that CTC patients who did not die during CTC or in hospital had the same annual mortality rate as all children aged under 5 in Zambia. The under five mortality rate in Zambia in 2006 was 182 per thousand live births [19]. Therefore we assumed that 3.64% (0.182/5) of children who recovered during CTC would die within a year.

Model Parameters: Cost of CTC and Hospital Care

All costs were expressed in international dollars for the year 2008. Unit costs measured in Zambian kwacha and UK pounds were deflated to their year 2000 equivalents [20,21], then converted to year 2000 international dollars using WHO exchange rates to reflect purchasing power parity [22]. They were then adjusted to year 2008 values using United States inflation rates from 2000 to 2008 [23].

The relevant types of cost were for health centre visits, RUTF, hospital admissions and Valid International's contribution to establishing the programme. Costs per health centre visit were based on the 2008 LDHMT budget, minus the proportion of the LDHMT budget devoted to non-health centre services (and the proportional administration costs), plus LDHMT staff salaries paid by the provincial health department. This total annual cost of health centre care was then divided by the number of health centre visits during 2007 to produce an average cost per health centre visit. The cost of community mobilisation was estimated from the 2008 LDHMT budget for community based child health activities, 10% of which was assumed to be for CTC (personal communication, Dr C Mbwili, LDHMT). This was multiplied by 2.4 years of the CTC programme and divided by the 3358 children treated, producing a mean cost of community mobilisation per child. The cost per kilogram of RUTF in Zambia was estimated by Valid International. Programme data showed that the mean body weight per child was 7.4 (standard error 0.065) Kg, and that each child received 200 kcal/Kg/day, which is equivalent to 1.9 (standard error 0.016) Kg RUTF per week. Costs of ambulatory CTC were health centre unit costs plus RUTF costs, multiplied by the duration of treatment. Treatment duration was stratified by CTC outcome (Table 1).

Valid International expenditure on the Zambia programme was reported from April 2005 to January 2008. This excluded RUTF production (which was already accounted for), and included administration, training, research, local and international travel, and consultancy. For each line item of expenditure, the proportion attributable to CTC was estimated by two senior Valid International personnel. This was divided by the number of children who received CTC over the same period, to produce an average cost to Valid International per child, regardless of duration of treatment.

Costs per day of hospital inpatient care at the Lusaka University Teaching Hospital were not available and so were based on WHO estimates of tertiary hospital care in Zambia, adjusted to include drug costs [24]. For children referred to hospital, these daily costs were multiplied by the average length of stay in the Lusaka University Teaching Hospital acute admission ward (personal communication, Dr B Amadi).

Analysis

The cost effectiveness analysis was carried out with Tree Age Pro Healthcare software and checked with Microsoft Excel. Point estimates were calculated for costs, outcomes, CTC effect (that is, difference in mortality) and incremental cost effectiveness ratios (CTC costs divided by CTC effect), using the point estimates for each model parameter.

Probabilistic sensitivity analyses [14,25] were conducted with Tree Age Pro Healthcare, to quantify the combined uncertainty about costs, effects, and cost effectiveness, based on the uncertainty about all of the model's parameters. First, the distribution of each parameter was defined (Table 1). Parameter standard errors were assumed to have normal distributions and were estimated from programme data where available. If not available, to be consistent standard errors were defined as 25% of the mean, so that 95% confidence intervals would be 50% more or less than the mean; the only exception was for mortality among defaulters, for which the standard error was larger to reflect greater uncertainty. Monte Carlo simulation was performed, with 10000 iterations per analysis. From simulation results we estimated 95% confidence intervals for each output of the model from their percentiles. We also calculated the probability that the intervention was cost effective for a range of values that society might be willing to pay to obtain one unit of effect (that is, dollars per life saved or per DALY gained).

Finally, one- and two-way sensitivity analyses were conducted. One way sensitivity analyses were calculated using the mean values of each parameter, and varying the values of one parameter at a time. Two way sensitivity analyses were conducted to examine the effects of simultaneously varying the values of two parameters.

Patient's consent and research ethics committee approval were not necessary because the study was based on aggregate programme data and published literature and did not require access to individual patient records.

Results

CTC cost an average of $203 per child, 70% of which was due to RUTF and Valid International's costs (Tables 2 and 3). Health centre visits and hospital admissions accounted for 30% of the CTC costs. Of Valid International's costs, 51% were for personnel, 42% were for travel and subsistence, and 7% were for other items.

The results of the Monte Carlo simulation, from which confidence intervals and probabilities of cost effectiveness were estimate, are shown in Figure 2. Each dot represents the cost and effect of CTC for each iteration. The expected mortality rates after one year were 9.2% with CTC and 20.7% with no treatment—a risk difference of 11.5% (Table 3). Thus one life would be saved for every 8.7 children who received CTC. The average increase in expected DALYs with CTC was 3.8 per child. The relative risk of death with CTC compared to doing nothing was 0.44 (95% CI 0.26–0.95).

Table 2. Mean costs of community-based therapeutic care per child

Cost item	Unit cost ($)	Mean number of items per child	Mean cost per child ($)	% of total
RUTF (Kg)	6.20	11.70	72.52	35.8
Technical support	68.69	1.00	68.69	33.9
Hospital per day	41.35	0.83	34.16	16.9
Health centre visits	4.24	6.16	26.10	12.9
Community mobilisation	0.66	1.0	1.06	0.5
Total			202.53	100.0

RUTF ready to use therapeutic food

Table 3. Costs and effects of community-based therapeutic care compared to no treatment

	CTC		No treatment		Difference	
	Mean	(95% CI)	Mean	(95% CI)	Mean	(95% CI)
Mean cost per patient ($)	203	(139–274)	0	0	203	(139–274)
Death rate (%)	9.2	(4.3–7.25)	20.8	(10.5–31.8)	11.5	(0.4–23.0)
Expected DALYs[a]	30.2	(29.3–31.2)	26.4	(22.7–29.8)	3.8	(0.14–7.7)

CTC community-based therapeutic care.
[a] assuming 33.3 disability adjusted life years (DALYs) expected after surviving one year [5]

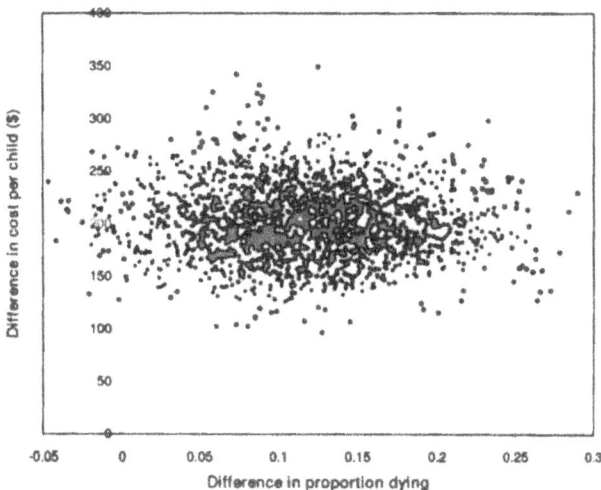

Figure 2. Incremental costs and effects from Monte Carlo simulation.

The cost of CTC was $1760 (95% CI $592–$10142) per life saved and $ 53 (95% CI $18–$306) per DALY gained.

CTC was more likely than not to be cost effective if society was willing to pay at least $1700 per life year gained (Figure 3). CTC was more than 80% likely to be cost effective if society was willing to pay at least $3000 per life saved. With regard to DALYs, CTC was more likely than not to be cost effective if society was willing to pay at least $52 per DALY gained (Figure 4). CTC was more than 80% likely to be cost effective if society was willing to pay at least $88 per DALY gained.

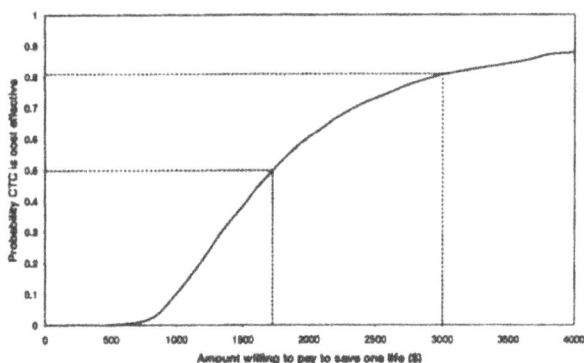

Figure 3. Probability CTC was cost effective for different amounts willing to pay per life saved.

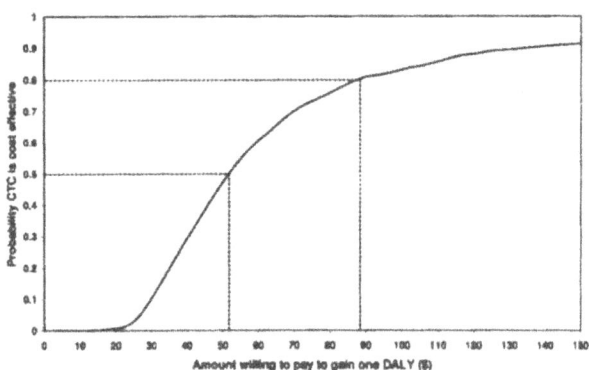

Figure 4. Probability CTC was cost effective for different amounts willing to pay per DALY gained.

The model was most sensitive to assumptions about expected mortality without treatment, weeks of CTC per child, effect of HIV on mortality without CTC, hospital referral rate, cost per kilogram of RUTF, quantity of RUTF consumed per week and technical support costs (Table 4). Cost effectiveness estimates were less sensitive to assumed unit costs of health centre visits and hospital admissions. The CTC outcome parameter which was least well known—death rates among defaulters—had relatively little influence on cost effectiveness estimates. The model's sensitivity to combinations of the most influential variables is shown in Figures 5 and 6. They show that the cost per life saved increased exponentially as the assumed death rate without treatment decreased towards 12%, and increased linearly with increasing weeks of CTC per child and cost per kilogram of RUTF.

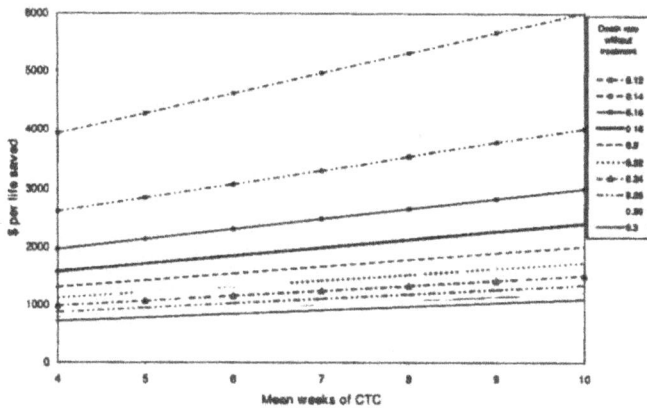

Figure 5. Cost per life saved for different assumptions about death rates without treatment and number of weeks of CTC per child.

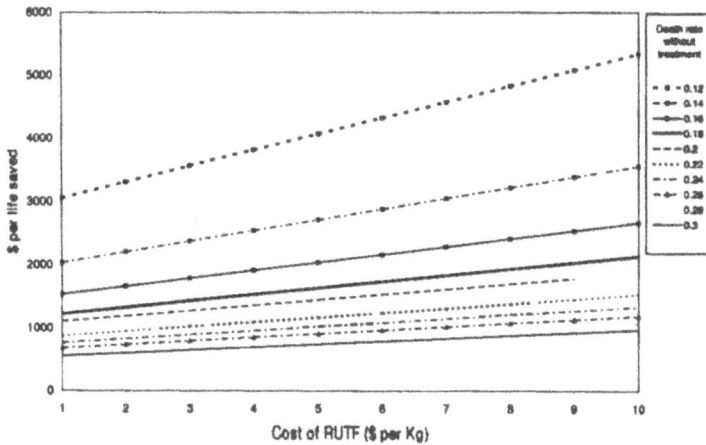

Figure 6. Cost per life saved for different assumptions about death rates without treatment and costs per kilogram of RUTF.

Table 4. Sensitivity analysis: cost per life saved for different values of model parameters

	Parameter values			Incremental cost effectiveness ratio ($ per life saved)		
	Base	Minus 50%	Plus 50%	Low	High	Range
Outcome probabilities						
Death rate within a year without CTC HIV- *	0.18	0.09**	0.27	17502**	927	16576
Relative risk of death if HIV+, no CTC *	2	1	3	2300	1426	874
Refer to hospital from CTC	0.059	0.030	0.089	1479	2100	622
Death within a year if recover with CTC	0.0364	0.0182	0.0546	1520	2091	571
HIV prevalence*	0.15	0.075	0.225	1994	1575	419
Death during CTC	0.026	0.013	0.039	1586	1978	392
Death rate in hospital	0.37	0.185	0.555	1608	1944	337
Death rate among defaulters	0.045	0.022	0.067	1703	1821	118
Default from CTC	0.173	0.087	0.260	1721	1802	82
Costs						
Weeks of CTC – Recovered	6.6	3.3	9.9	1419	2101	682
RUTF per Kg ($)	6.20	3.10	9.30	1445	2075	630
Mean Kg of RUTF per child per week	1.90	0.95	2.85	1445	2075	630
Valid cost per child ($)	68.69	34	103	1462	2058	597
Hospital cost per day ($)	41.35	20.68	62.03	1612	1908	297
Days in hospital	14	7	21	1612	1908	297
Costs per health centre visit ($)	4.24	2.12	6.36	1647	1873	227
Weeks of CTC – Defaulted	5.1	2.6	7.7	1699	1821	123
Weeks of CTC – Referred	4.8	2.4	7.2	1740	1780	39
Weeks of CTC – Died	3.6	1.8	5.4	1754	1767	13
Community mobilisation per child $	1.06	0.5	1.6	1755	1765	9

* CTC is more cost effective if these parameters are greater.
** note higher mortality with CTC than with no care

Discussion

The study shows that CTC for SAM among children aged under five years in Lusaka results in good outcomes at a reasonable cost. The estimated cost of $1760 per life saved, or $53 per DALY gained, suggests that this model of CTC has a similar cost effectiveness to other priority child health interventions in Africa such as immunisation, micronutrient supplementation, and treatment of pneumonia and diarrhoea [26]. The cost per DALY gained was very similar to the result of a World Bank study based in Guinea in 1998 [27], despite the different methods of evaluation. WHO has classified child health interventions as highly cost effective if the cost per DALY gained is less than the country's gross national product per capita [15,26]. This can be used as one indicator of society's willingness (or at least ability) to pay for improved health. This supports CTC, since Zambia's gross national income per person per year was $1000 in 2006 [28].

The main strengths of this study are that 1) it was based on an innovative large scale programme implemented through government primary care health centres throughout a Zambian city, 2) it had original and up to date programme data on costs, outcomes and severity of SAM and 3) it compared the costs and outcomes of the programme with what would be expected without any intervention. Comparison with no treatment allows the cost effectiveness of this model of CTC to be compared with any other intervention in health, and not just to be considered as an incremental change to alternative nutrition strategies [15,29]. This method of comparing a health care intervention with the hypothetical alternative of doing nothing, and using probabilistic sensitivity analyses, follows WHO health economists' recommendations for economic evaluation and priority setting [15,30].

The main limitation of comparing CTC to doing nothing is that it is dependent on assumptions about the outcomes and costs of no health care for SAM. We do not know what these children's mortality rates would have been without treatment. Most recent studies of mortality among children with SAM have been among children who received treatment [5,9,10]. We therefore relied on population based cohort studies in Africa carried out about 20 years ago when little health and nutritional care was available [1-4]. However, we have accounted for the likely effect of the HIV epidemic, and UNICEF estimates indicate that under five mortality rates in Zambia remained constant between 1990 and 2006 [31]. Although comparable mortality rates were not available for children with oedema, this programme's data showed that mortality among children with oedema was the same as for children with MUAC of 11 cm or less and with no oedema. Our assumption of 18% mortality without treatment does not seem excessive considering that, in 9 randomised trials of hospital treatment of SAM, short term mortality rates ranged from 16%–46% (median 20%) among control group children who received conventional treatment [5]. Recent African hospital case series have also shown higher short term mortality rates for children with MUAC < 11.5 cm, despite treatment [18,32,33]. Furthermore, UNICEF estimates [31] show that under five mortality rates in Zambia were higher than in Malawi and Uganda where the cohort studies were conducted. Even if we assumed that mortality rates without treatment were as low as 12%, CTC was still relatively cost effective, with a cost per life saved of about $5000 (Figure 5), and with a cost per DALY gained of $150.

The deterministic sensitivity analyses (Table 4) showed that results were most sensitive to assumptions about mortality without treatment (discussed above), costs of RUTF, and costs of technical support, which has implications for future research priorities. The importance of mortality without CTC highlights the need for future studies to track mortality in comparator populations. RUTF accounted for 36% of total CTC costs, underlining the desirability of reducing RUTF costs in future. It seems likely that an alternative less costly RUTF formulation, using locally grown soya, sorghum and maize instead of imported milk powder and peanuts, could be as effective at lower cost. Larger scale production and food procurement would also be likely to reduce costs in future. This alternative will be evaluated in a randomised trial and economic evaluation soon to be started in this setting. The relatively high cost per child of Valid International's input was largely due to technical support while setting up the programme. At least some of these costs could thus have been considered as capital costs and spread over a longer period, as the programme becomes increasingly run by government health services. In future Valid International's emphasis will change from ongoing technical support to training health ministry trainers, which could be less costly. Thus costing of longer term and larger scale implementation will be needed in future.

Several other limitations and assumptions of the study should be considered. First, it would have been desirable also to have compared CTC with alternative ways of treating SAM. As stated in the Background, although other studies have compared outcomes of community- and hospital-based treatment of SAM [11,12], only one study compared costs [13]. If community based care was at least as effective as hospital-based care [13], it is plausible that the former would be less costly and thus more cost effective. However we could not make such a comparison with local data because our examination of clinic records found that comparable outcome measures prior to CTC were not available. Second, unit costs of hospital and health centre care were not available specifically for children with SAM. However neither of these unit costs had much influence on overall cost and cost effectiveness estimates (Table 4). Third, the mortality rate among children who defaulted was unknown but this too had relatively little influence. Fourth, HIV prevalence among this population was not known. If HIV prevalence was higher than the 15% assumed, that would increase the estimated cost effectiveness of CTC by increasing the mortality rate without CTC. Fifth, DALY estimates assumed that health and life expectancy would return to normal after recovery, although it is plausible that children who recovered would be at higher risk of stunting and poorer health in future [5]. Alternatively, they could be relatively hardy survivors. However, it is easy for readers to adjust these effectiveness and cost effectiveness estimates to reflect different assumptions about expected DALYs after recovery. For example, if one assumed that 25 DALYs were expected after recovery, instead of 33.3, the cost per DALY gained would be $53 × 33.3/25 = $71. Finally, discounting was not used because child level costs covered less a year and expected DALYS were assumed to be net present values.

Conclusion

The Lusaka model of CTC for SAM appears highly cost effective. This study suggests that this form of CTC should be expanded to the rest of Zambia and adapted for other African countries with high rates of SAM. Cost effectiveness could be increased in future with less external technical support, as CTC is increasingly implemented through government services, and by reducing RUTF costs through local and larger scale production and sourcing of components. Priorities for future research include controlled trials and economic evaluations of alternative ways of providing CTC, such as hospital-based care or selective SAM-only programmes.

This requires prospective collection of individual level data on severity of SAM, HIV status, use of health services and outcomes, and active follow up of children who default.

Abbreviations

CTC: community-based therapeutic care; LDHMT: Lusaka District Health Management Team; MUAC: mid upper arm circumference; RUTF: ready to use therapeutic food; SAM: severe acute malnutrition; UNICEF: United Nations Children's Fund; WHO: Word Health Organization,

Competing Interests

The author received funding from Valid International and Concern to carry out the study.

Authors' Contributions

MOB designed the model, identified relevant parameters and data sources, carried out the analyses and wrote the paper.

Authors' Information

Max Bachmann is a health services researcher and public health physician who uses health economic and clinical epidemiological methods to evaluate innovative health care interventions to improve population health, such as child health care, HIV/AIDS care and chronic disease management, in Africa and the United Kingdom. Now at the University of East Anglia, he previously worked at the medical schools of the universities of Cape Town, Bristol and the Free State.

Acknowledgements

Clara Mbwili Muleya provided LDHMT financial data, Beatrice Amadi provided information on Lusaka University Teaching Hospital; Valid International personnel contributed as follows: Paul Binns provided RUTF costs and comments on the paper, Prosper Dibidibi Kabi and Abel Hailu provided programme data, Victor Owino facilitated local contacts in Lusaka, Valbona Luci reported Valid

International's expenditure, Alistair Hallam provided cohort study literature, and Steve Collins and Paluku Bahwere commissioned the study, apportioned Valid International's costs and commented on the paper; Nicky Dent of Concern commented on the paper; and Valid International and Concern funded the study. I am grateful to the journal's two reviewers for their suggestions.

References

1. Pelletier DL: The relationship between child anthropometry and mortality in developing countries: implications for policy programs and future research. J Nutr 1994, 124(10 Suppl):2047S-2081S.

2. Pelletier DL, Low JW, Johnson C, Msukwa LA: Child anthropometry and mortality in Malawi: testing for effect modification by age and length of follow-up and confounding by socioeconomic factors. J Nutr 1994, 124:2082S-2105S.

3. Vella V, Tomkins A, Ndiku J, Marshal T, Cortinovis I: Anthropometry as a predictor for mortality among Ugandan children, allowing for socio-economic variables. Eur J Clin Nutr 1994, 48:189–97.

4. Vella V, Tomkins A, Borghesi A, Migliori GB, Ndiku J, Adriko BC: Anthropometry and childhood mortality in northwest and southwest Uganda. Am J Publ Health 1993, 83:1616–8.

5. Bhutta Z, Ahmed T, Black RE, Cousens S, Dewey K, Giugliani E, Haider BA, Kirkwood B, Morris SS, Sachdev HPS, Shekar M, Maternal and Child Undernutrition Study Group: What works? Interventions for maternal and child undernutrition and survival. Lancet 2008, 371:417–440.

6. Briend A, Lacsala R, Prudhon C, Mounier B, Grellety Y, Golden MH: Ready-to-use therapeutic food for treatment of marasmus. Lancet 1999, 353:1767–1768.

7. Briend A: Highly nutrient-dense spreads: a new approach to delivering multiple micronutrients to high-risk groups. Br J Nutr 2001, 85(Suppl 2):S175-S179.

8. Bahwere P, Binns P, Collins S, Dent N, Guerrero S, Hallam A, Khara T, Lee J, Mollison S, Myatt M, Saboyo M, Sadler K, Walsh A: Community Based Therapeutic Care. A Field Manual. Oxford, Valid International; 2006.

9. Collins S, Dent N, Binns P, Bahwere P, Sadler K, Hallam A: Management of severe acute malnutrition in children. Lancet 2006, 368:1992–2000.

10. Ashworth A: Efficacy and effectiveness of community-based treatment of severe malnutrition. Food Nutr Bull 2006, 27(Suppl):S24-S48.

11. Ciliberto MA, Sandige H, Ndekha MJ, Ndekha MJ, Ashorn P, Briend A, Ciliberto HM, Manary MJ: Comparison of home-based therapy with ready-to-use therapy with standard therapy in the treatment of malnourished Malawian children: a controlled, health centre effectiveness trial. Am J Clin Nutr 2005, 81:864–870.

12. Patel MP, Sandige HL, Ndekha MJ, Briend A, Ashorn P, Manary MJ: Supplemental feeding with ready-to-use therapeutic food in Malawian children at risk of malnutrition. J Health Pop Nutr 2005, 23:351–357.

13. Ashworth A, Khanum S: Cost-effective treatment for severely malnourished children: what it the best approach? Health Policy Plan 1997, 12:115–121.

14. Drummond MF, Sculpher MJ, Torrance GW, O'Brien B, Stoddart GL: Methods for the Economic Evaluation of Health Care Programmes. 3rd edition. Oxford, Oxford University Press; 2005.

15. Tan-Torres Edejer T, Baltussen R, Adam T, Hutubessy R, Acharya A, Evans DB, Murray CJL, (Eds): Making Choices in Health: WHO Guide To Cost-Effectiveness Analysis. Geneva, WHO; 2003.

16. UNAIDS/WHO: Global HIV/AIDS Online Database. 2006 Report on the Global AIDS Epidemic. [http://www.unaids.org]

17. Central Statistics Office Zambia. Census Data from Zambia [http://www.zamstats.gov.zm/census.php]

18. Amadi B, Kelly P, Mwiya M, Mulwazi E, Sianongo S, Changwe F, Thomson M, Hachungula J, Watuka A, Walker-Smith J, Chintu C: Intestinal and systemic infection, HIV, and mortality in Zambian children with persistent diarrhoea and malnutrition. J Ped Gastr Nutr 2001, 32:550–554.

19. UNICEF: State of the World's Children 2008. [http://www.unicef.org/publications/index_42623.html] Geneva: UNICEF; 2008.

20. Bank of Zambia. Snapshot Inflation [http://www.boz.zm/snapshot_inflation.htm]

21. Bank of England. Inflation Calculator [http://www.bankofengland.co.uk/education/inflation/calculator/flash/index.htm]

22. WHO-CHOICE. CHOosing Interventions that are Cost Effective [http://www.who.int/choice/costs/en/]

23. U.S. Department of Labor. Bureau of Labor Statistics. Consumer Price Index [http://www.bls.gov/CPI/]

24. Adam T, Evans DB, Murray CJ: Econometric estimation of country-specific hospital costs. Cost Effect Res Allocation 2003, 1:3.

25. Ades AE, Claxton K, Sculpher M: Evidence synthesis, parameter correlation and probabilistic sensitivity analysis. Health Econs 2006, 15:373–381.

26. Tan-Torres Edejer T, Aikins M, Black R, Wolfson L, Hutubessy R, Evans DB: Cost effectiveness analysis of strategies for child health in developing countries. BMJ 2005, 331:1177. |

27. Prabhat JHA, Bangohoura O, Ranson K: The cost-effectiveness of forty health interventions in Guinea. Health Policy Plan 1998, 13:249–262.

28. World Bank: World Development Report 2008. Washington: World Bank; 2008.

29. Hutubessy R, Chisholm D, Tan-Torres Edejer T, WHO-CHOICE: Generalized cost-effectiveness analysis for national-level priority setting in the health sector. Cost Effect Res Allocation 2003, 1:8.

30. Baltussen RMPM, Hutubessy RCW, Evans DB, Murray CLJ: Uncertainty in cost-effectiveness analysis: probabilistic uncertainty analysis and stochastic league tables. Int J Technol Assess Health 2002, 18:112–119.

31. UNICEF: The State of the World's Children 2008. Child Survival. New York: UNICEF; 2007.

32. Berkley J, Mwangi I, Griffiths K, Ahmed I, Mithwani S, English M, Newton C, Maitland K: Assessment of severe malnutrition among hospitalized children in rural Kenya: comparison of weight for height and mid upper arm circumference. JAMA 2005, 294:591–597.

33. Bitwe R, Dramaix M, Hennart P: Simplified prognostic model of overall intrahospital mortality of children in central Africa. Trop Med Int Health 2006, 11:73–80.

Integrating Chronic Care and Business Strategies in the Safety Net: A Practice Coaching Manual

Katie Coleman, Marjorie Pearson and Shinyi Wu

Introduction

Welcome! Most likely, if you're reading this you are interested in improving health care quality through practice coaching. This practice coaching manual aims to help effectively and efficiently improve clinical quality in an ambulatory setting by providing:

- An OVERVIEW of what practice coaching is and how a variety of settings have used it to improve care.

- A SUMMARY of important characteristics and skills to look for when recruiting or training a practice coach.

- A DESCRIPTION of a time-limited practice coaching intervention that includes a series of activities, companion agendas, and tools.

This practice coaching manual accompanies a comprehensive Web-based toolkit, "Integrating Chronic Care and Business Strategies in the Safety Net." The toolkit outlines a sequence of steps that practice teams can use to efficiently improve clinical quality along the lines of the Chronic Care Model. It also includes presentations, assessments, data tracking sheets, and sample action plans for use by teams as they transform their care. The toolkit and this practice coaching manual work together and refer to each other. We know that clinical teams often need help and support to effectively improve care, and we believe practice coaching may be useful to them as they do this work. This manual provides instructions and materials needed to support those using "Integrating Chronic Care and Business Strategies in the Safety Net" to transform care.

The development of these two resources grew out of a desire to help primary care teams improve clinical quality efficiently and effectively. Both are based on the Chronic Care Model (CCM), an evidence-based framework that has helped hundreds of clinical practices transform their daily care. The Chronic Care Model (CCM) is designed to help practices improve patient health outcomes through changing the routine delivery of ambulatory care. The Model calls for a number of interrelated system changes, including a combination of effective team care and planned interactions; self-management support bolstered by more effective use of community resources; integrated decision support; and patient registries and other supportive information technology.

Most often, the CCM has been implemented through Breakthrough Series (BTS) Collaboratives, an organized quality improvement approach that brings together practices from a variety of organizations four times a year to learn from leaders and colleagues about improving care. In between these learning sessions, teams return to their practices and try out new ways of delivering care through small, short-cycle changes called Plan-Do-Study-Act (PDSA) cycles. The practices that have participated in BTS Collaboratives to learn the CCM improved the care they provided for patients and improved patient health outcomes.[1-8]

In our 10 years of experience with BTS Collaboratives, we have seen that they are often expensive to organize and require practices to take time out from providing patient care to attend learning sessions. Often the practices that are willing and able to do this are more highly motivated and well-supported than others. We sensed a need for a less time- and resource-intensive intervention that would:

- Make the tools and concepts taught in the Collaboratives available to more practices, and

- More closely integrate the business strategies necessary to sustain clinical change in the long term.

The manual is created primarily based on our practice coaching experience during the AHRQ-funded pilot project "Integrating Chronic Care and Business Strategies in the Safety-Net." It captures our coaches' approach to the teams, lessons learned from our experience, and feedback from the teams. It is supplemented by a literature review and interviews with leaders from other national coaching initiatives.

This Practice Coaching Manual Is Designed for:

- Clinic or hospital leaders who want to use coaching to initiate or spread improvement efforts from one site to others;
- Quality improvement coaches, improvement leaders, and anyone else interested in new ideas about how to facilitate practice improvement; and
- Public health departments, multistakeholder collaboratives, and medical associations or other organizations interested in improving clinical quality in medical practices.

This practice coaching manual and the companion toolkit are meant to provide the tools and structure for coaches to use in helping teams in a wide variety of settings improve clinical quality. Of course, modifications and tailoring for the specific context where you work may be appropriate. However, many of the tools in the companion toolkit are copyrighted and cannot be modified unless the original authors grant permission. As this is an emerging coaching model, we would love to hear from you about your experience using this manual and toolkit.

What is Practice Coaching?

In 2006, the Agency for Healthcare Research and Quality funded a project to develop, test, and disseminate a package of tools to facilitate the effective and financially viable implementation of the Chronic Care Model (CCM) in safety net organizations. The RAND Corporation, Group Health's MacColl Institute for Healthcare Innovation, and the California Health Care Safety Net Institute participated in the project. A key premise of our effort was that primary care practices may need more help than a toolkit alone can provide, yet they may be unable to attend a year-long Breakthrough Series style collaborative. This intervention was designed to provide low-intensity in-person, hands-on guidance to successfully implement the CCM. We conceptualized such assistance as helping, advising, and enabling and used terms such as "coaching" and "facilitation" when talking about it.

To better understand how such help might be structured, we looked at the literature on coaching and facilitation and talked with nine practice coaching leaders from a variety of organizations. In this chapter, we summarize key lessons learned from the literature and our interviews with coaching practitioners, as well as our own experience with practice coaching.

Why Practice Coaching?

There are a number of reasons that primary care organizations might want to look to coaches when embarking on a program of practice improvement:

- Primary care practices often lack in-house expertise or experience to successfully identify and initiate needed changes. Coaches can bring expertise on specific topics and approaches, and tools to facilitate implementation.

- Practice transformation is a complex undertaking, involving fundamental change to how a practice operates. Coaches have experience in how to help practices sequence and manage change.

- Primary care practices have difficulty making time for quality improvement in the face of the competing demands of day-to-day practice. The presence of a coach lends structure, dedicated time, and focus to quality improvement efforts.

What Roles Do Practice Coaches Play?

Coaches perform multiple functions.[9] Coaches can serve as:

- Facilitators who help practices achieve their improvement goals.
- Conveners who bring groups of staff members together to work through an issue.
- Agenda setters and task masters who help practices prioritize their change activities and keep them on track.
- Skill builders who train practices in quality improvement processes and assist them in developing proficiency in the techniques used in the CCM.
- Knowledge brokers who know about external resources and tools and save practices from engaging in extensive searches for information or reinventing the wheel.
- Sounding boards who give practices a reality check and provide feedback.

- Problem-solvers who can help practices identify and surmount a stumbling block.
- Change agents who promote adoption of specific evidence-based practices.

What Do Practice Coaches Do?

Coaches can play a role in setting the stage at the outset of the transformation process. For example, coaches can:

- Help to prepare the organizational infrastructure for quality improvement implementation through such activities as advising on team-building, improving communication,[10] facilitating meetings,[11] and helping to develop leadership skills.[12]

Communicate the vision for change through activities such as presenting best practices[13-15] and sharing what other organizations have done.

- Help people to better understand how their practice compares to the ideal and where there is room for improvement by observing and delineating practice operations, assessing needs, and gathering baseline data, as well as guiding discussions of the current practice and opportunities for change.[16, 17]

Coaches can also engage in very concrete tasks during the implementation period. Coaches can:

- Help practitioners to plan change by encouraging them to set goals,[18] suggesting ideas or providing menus of possible strategies or innovations,[16, 17] and helping them choose among such options and create a plan.[15-17]
- Enable practitioners to execute changes by providing tools,[13, 14, 16, 17, 19] guiding them through rapid-cycle tests of change,[11, 13, 14, 18] and assisting when obstacles arise.[11]
- Aid practices in customizing processes to fit their own situation and incorporating the changes into their day-to-day routines, so as to increase the likelihood that the changes will be sustained.[20, 21]
- Provide direct technical support with health information technology (HIT) implementation and development of registries and reminders systems.[15]
- Help practitioners to collect and use measurement data,[22] assess the effectiveness of changes made[16, 17] and sometimes even undertake activities such as conducting chart audits.[15-17]

Motivation, education, and consultation are at the core of coaching.

- Motivational coaching addresses the amount of effort that group members collectively put into the task, especially by enhancing the conviction and confidence they bring to the work[23] through encouragement, reassurance, permission, and nudges.[24]

- Educational coaching addresses the knowledge and skills that members bring to bear on the group's work.[23] Educational coaching can take the form of information sharing, skills training, and role feedback.[24]

- Consultative coaching fosters use of performance strategies that are especially well-aligned with and appropriate to the task.[23] Consultative coaching may include rapid response to needs and requests; interactive problem solving,[17, 24] and suggestions for change concepts or resources.

Most coaching involves a mix of these functions, but the emphasis placed on any one function changes over the course of the coaching process.[9, 23] A motivational focus, for example, may be needed before education or consultation can be effective.

A frequent challenge for coaches is to maintain clarity about what they do and do not do. Coaching leaders have observed that there is a danger of "scope creep," whereby coaches are pulled into work unrelated to the project at hand. In most cases this occurred because the coaches themselves were not clear on their role or because they wanted to be perceived as a helpful and valuable resource. "Scope creep" was best managed through clarification of roles at the outset of the project, frequent reevaluations of project status and open, clear communication with both the practice team and their leadership about the role of the coach and the expectations of the teams.

How Is Practice Coaching Structured?

Coaching approaches and methods vary in many respects, including:

- Duration (e.g., from a few months to a number of years).

- Intensity, ranging from time-intensive, comprehensive practice management and clinical quality improvement efforts involving frequent communication with sites (e.g., ongoing facilitation provided through practice-based research networks) to brief and narrowly focused efforts (e.g., a preventive care effort launched with one group meeting and minimal follow-up).

- Proximity, ranging from onsite coaching, with a coach dedicated to a single site or set of sites (e.g., academic research institute coaches integrated into university-affiliated practices) to long-distance coaching, using telephone and e-mail to continue work between in-person meetings (e.g., coaches in large systems such as the Veterans Affairs (VA) health system).

Coaching also can be:

- A team activity, whereby two or more coaches bring complementary skills to interactions with the practice (e.g., specialized expertise in improvement methods versus the clinical problem area).

- Scripted, using a consistent curriculum for practice coaches to use with sites (e.g., Improving Performance in Practice).

- Prescriptive of the changes that the practices should make (e.g., top-down promotion of highly defined best practices).

- Practice driven, allowing the structure—and to some degree the content—of the program to be decided largely by the site (e.g., STEP-UP16).

Most coaching leaders acknowledged a tension between wanting to be reliable and consistent in their approach to teams while recognizing that one key advantage of coaching is the ability to tailor the implementation of a quality improvement initiative to needs and strengths of each practice. Learning which elements of an intervention work and are generalizable and which can and should be customized at the site level is an area where much more needs to be known.

Who Serves as a Practice Coach?

While coaching can be done by a member of the practice, the predominant model found in the literature is to use a coach external to the practice. In the coaching interventions that we studied, an entity outside the practice arranged and paid for the coaching. Practice coaching is a service available for purchase. A variety of different types of individuals have served as coaches. These include:

- Researchers with expertise on evidence-based practice and implementation (e.g., practice facilitators for the VA Quality Enhancement Research Initiative (QUERI) program).[24-27]

- Professional improvement advisors, broadly trained in quality improvement methods (e.g., faculty at the Institute for Healthcare Improvement).

- Specially trained individuals with bachelor's or master's degrees and some previous health care experience or training (e.g., practice enhancement assistants trained by practice-based research networks).

Which Practices Benefit from Coaching?

It is difficult to predict which practices will be most likely to succeed. Coaches generally see that practices with engaged leaders and long-term quality improvement goals are more likely to embrace the changes coaches nurture. On the other hand, programs using coaches may want to target practices unlikely to be able to engage in quality improvement on their own. These include practices that:

- Are not part of or supported by a larger system.
- Cannot attend quality improvement collaboratives.
- Require additional motivation or contain pockets of resistance or inertia that block spread of the CCM.

Does Practice Coaching Work?

Although there are few evaluations of practice coaching, it is perceived to be valuable. Many have come to view primary care practices as complex adaptive systems, each with unique histories, people, relationships, values, rules, influences, and problems.[28, 29] Since one predefined approach cannot possibly fit all these unique systems, quality improvement implementation requires extensive customization. This customization, in turn, necessitates understanding the context and opportunities for change[30] and facilitating a process of learning and reflection that helps practices adapt to and plan change.[31] Coaching is key to this process.

Emerging evidence suggests that this tailoring to the practice's unique context may increase the likelihood of sustainability by helping to better incorporate quality improvement changes into the day-to-day routines of the practice.[20, 21] Studies have shown that coaching has led to increases in evidence-based care of diabetics, preventive services, and screenings.[15, 32, 33]

Evaluation of our practice coaching intervention, which was designed to foster adoption of the CCM and use of the "Integrating Chronic Care and Business Strategies in the Safety Net" toolkit, has led to the following conclusions:

1. *Coaching is a necessary bridge to the toolkit.* The coaches help providers and staff navigate the toolkit. By answering questions and helping people locate specific tools, the coaches save staff and provider time.

2. *Coaching motivates and prompts people to make changes.* The coaches encourage providers and staff to test small changes in their work routines, which providers and staff may not have been able to do on their own. The participants believed these changes would not have happened without coaching.

3. *Coaching extends the horizons of the teams.* The coaches provided outside experience and shared information from other clinics. These examples allowed the providers and staff to learn from changes that have been effective elsewhere, resulting in greater motivation in implementing the CCM.

4. *Coaching has a positive effect on team building.* Although some physicians and their supporting staff worked well together prior to the project, others commented that coaching helped them to build a better team through regular meetings and staff empowerment.

5. *Coaching is an emotional bond.* The coaches' commitment and positive attitudes in motivating and encouraging participants were appreciated. This emotional bond was noted to be a key factor in the success of the coaching intervention.

What Makes a Good Practice Coach?

For those practices interested in hiring their own practice coach, below are some characteristics to consider, including a list of core competencies and a proposed scope of work. Because this area has not been empirically examined in the context of ambulatory care, we rely on our own experience and our conversations with national leaders to suggest what makes a good practice coach.

Characteristics

In our experience and that of others in the coaching world, certain characteristics and personality traits of the coach are tremendously important. Because of the interpersonal nature of the coaching relationship, respect for others, superior communication skills, and open-mindedness are characteristics deemed most crucial. Other characteristics mentioned by experienced program leaders as important for a potential coach include empathy, creativity, passion for the job, and respect for the real-life barriers in practice. They also need to have a thick skin and avoid internalizing things. Being a "people-person" was considered very important (e.g., being able to get al.ong well with people and being good at reading people and understanding who is in power). Teaching skills also were emphasized, as was the ability to read between the lines and elicit underlying issues in a nonthreatening way.

Those quality improvement leaders who have experience serving as practice coaches spoke about some of the challenges of working with different types of people on different teams. The executive director of one quality improvement effort said, "Coaches must have a variety of approaches at their fingertips to connect with different teams. And, you need lots of different tools in your toolbox to connect with different types of staff—from those with a high school education to highly trained providers. A coach has to work well with all of them."

In many cases, the coach is the face of the quality improvement program for the practice teams. Being able to keep teams engaged in what is often very challenging improvement work is not easy. As one coach put it, "You have to have a thick skin. There is no way around it. You'll be treated like dirt, and you can't take it personally." Sometimes the frustration of the team gets directed at the coach, so being able to maintain good relationships while continuing to promote improvement is key.

Core Competencies

In addition to the interpersonal skills and emotional intelligence of coaches that may enable them to function well in a practice, some skills and content knowledge are needed. Although all our interviewees agreed that these skills were important, there was some debate as to which were essential and which were nice to have. If you are fortunate enough to have a number of coaches that will work together on your initiative, then the group as a whole could possess these skills. Each coach individually may be able to provide specialized knowledge in areas where they are more familiar. If you only are able to hire one coach, seeking out external sources of support in areas where that person may not be as strong would be helpful.

Skills and knowledge a coach should possess or be able to connect with include:

- Familiarity with data systems, including registries.
- Ability to understand and explain data reports in different ways to different stakeholders.
- Some clinical understanding and credibility.
- Knowledge of, and experience with, the Chronic Care Model.
- Knowledge of, and experience with, the Model for Improvement.
- Understanding of performance reporting and measurement.
- General quality improvement methods.
- Group facilitation skills.
- Project management skills.
- Knowledge of practice management and/or financial aspects of the practice.
- Experience with and understanding of the outpatient clinical setting.

There was considerable debate about how important it is for the coach to be clinically trained, such as a registered nurse, physician assistant, nurse practitioner, medical doctor, or doctor of osteopathy. Some thought it was essential that the coach be a clinician to provide credibility and to act as a resource with whom the practicing physician could discuss clinical issues in improvement. Others thought having a clinician coach may be a detriment because of an overemphasis on the clinical aspects of care. These respondents stressed the wide variety of skill sets needed to care for patients and emphasized how a coach needs to be able to value and speak to each role. In the end there are pros and cons to having a clinician coach. Likely it is important for the coach to have some clinical credibility and to be able to access a provider to come in and talk to the clinicians on an "as-needed" basis.

How Much does Coaching Cost?

There is little information about the costs of coaching, which of course varies with the intensity of coaching, the qualifications of the coach, and the duration of the coaching. Our 10-month practice coaching of two clinic sites cost approximately $41,000 (in 2007 dollars), which included time spent in coach training, coaching, and travel to sites. Practice coaching has been shown to be cost-effective by reducing inappropriate testing and treatment costs and increasing practice efficiency.[34]

How does Coaching Compare to Participating in a Collaborative?

More than 1,500 physician practices have participated in CCM collaboratives. Collaboratives can be thought of as group coaching sessions, where several practices are all trained in CCM implementation at the same time. There is real value in bringing together groups of practices. Teams benefit when they get together to interact, share lessons learned, feel some camaraderie with colleagues undergoing similar transformation, and develop ongoing networks.

Coaching, however, may be uniquely beneficial in these ways:

- Coaches can see and evaluate practice resources firsthand and tailor advice accordingly.
- Bringing coaches to the practice can enable more staff to participate in the practice improvement sessions.

- Through shorter educational sessions, conducted during a lunch break or after work hours, coaching can be delivered without requiring the closing of the practice.

Coaching has also been used as a supplement to collaborative learning sessions, blending the best that both methods have to offer.

Clearly the field of practice coaching is still evolving, and it may be that even as our knowledge base grows, different models will work better in different settings. The next chapter provides a detailed description of the practice coaching intervention developed to be used in tandem with the "Integrating Chronic Care and Business Strategies in the Safety Net" toolkit, available at www.ahrq.gov and www.improvingchroniccare.org.

Successful Coaching Case Study #1

Coaching Preventive Care Improvement in Primary Care Practices[34-36]

Who was coached?

Fifty-four physicians and allied health staff in 22 primary care practices in Ontario, Canada.

Who were the coaches?

Three "prevention facilitators," all nurses with community nursing degrees and previous facilitation experience. They received 30 weeks of training in outpatient medical systems and management, preventive improvement, performance reporting, and facilitation techniques. Each coach was assigned to up to eight practices (with up to six physicians per practice) within a geographic area.

How was the coaching structured? The coaches worked out of their homes and traveled by car to the practice locations for onsite visits. During the 18 months of the intervention, they made 33 visits to each practice and spent 1 hour and 40 minutes per visit, on average. Between visits, they corresponded regularly with each practice through e-mail and telephone calls.

What roles did coaches play?

The coaches served as educators, providing evidence on best preventive practices; motivators, using audit and feedback as well as opinion leader strategies; consultants, offering specific improvement tools and strategies such as reminder systems; team conveners and consensus builders; and chart auditors.

What did coaches do?

They presented baseline performance data; facilitated the meetings in which the practices set performance goals, developed prevention plans, and developed and adapted strategies and tools to implement these plans; and conducted chart audits to provide performance data to monitor success.

Successful Coaching Case Study #2

Coaching Local Development of Interventions to Improve Depression Recognition and Treatment in Substance Abuse Clinics[27]

Who was coached?

Clinicians and administrators at two outpatient substance abuse disorder clinics of the U.S. Department of Veterans Affairs (VA).

Who were the coaches? The facilitators, in this case, were researchers from the VA's Center for Mental Healthcare and Outcomes Research, including the project's principal investigator (PI), (a PhD with a background in sociology) and the project coordinator.

How was the coaching structured?

This facilitation used onsite visits, conference calls, site-specific diagnostic data, expert consultation, and provision of implementation strategies and tools to help the local teams design and launch the site-specific interventions to further adoption of guideline-based practices for recognizing and treating depression. The PI spent 16 hours per week and the project coordinator 30 to 40 hours on these diagnostic and design guidance activities.

What roles did coaches play?

The coaches acted as observers of local practice, collectors and providers of data and tools, educators on guideline-recommended practices, and builders of local expertise in quality improvement.

What did coaches do?

They used formative evaluation and local teams (called "Development Panels") to facilitate the development of the interventions. In the formative evaluation, the coaches used clinic observation and key informant interviews to diagnose the key facilitators of, barriers to, and influences on depression recognition and treatment in these clinics. Specific diagnostic activities of the coaches included (1) an initial visit to each clinic by the PI to review materials on policies and procedures and to meet with clinical directors, (2) a three-day visit to each clinic three months later to conduct formal and informal observations of program operations and to interview program staff (10 to 14 staff members at each site) and patients (five or six), and (3) an analysis and presentation of this information in tables that summarized problems and offered potential solutions and tools. Over the next 5 months, the coaches used conference call meetings to guide the local Development Panels (consisting of the clinical director, a physician, a counselor, and a nurse or other staff member involved in depression screening) in designing the intervention specifics for their clinic.

An Approach to Practice Coaching

In this chapter we describe the approach used by two coaches as they worked together with nine randomly selected primary care teams to improve quality of care. Feedback from the teams and reflections on how to alter and improve the intervention are also included. The described approach illustrates how the principles described in Chapter 2 were put into action. The goal of coaching was to lay the foundation for implementation of the CCM. This was done by tutoring practices in the CCM and quality improvement methods and acquainting them with the toolkit, which they could continue to use to guide their improvement activities after coaching ended. The tools and steps below provide a template for a practice coaching intervention, but organizations can and should adapt the pace and content of the work to fit their needs.

In a Nutshell

Who was coached?

Nine randomly selected primary care teams at two public hospital outpatient clinics located in California, USA. Both clinics were designated Federally Qualified Health Centers, serving disproportionately low-income and uninsured residents.

Who were the coaches?

Two quality improvement professionals external to the public hospital systems with expertise in teaching the Chronic Care Model and Model for Improvement and leading teams through quality improvement initiatives efficiently. Two coaches were used because of their complementary skill sets. One acted as the regular point of contact with teams. The other provided specific technical assistance around topics including selecting and monitoring performance measures, integrating self-management support into the routine visit, and developing and using registries.

How was coaching structured?

The coaching intervention was low intensity. The out-of-town coaches made two site visits and communicated with practices by phone two to three times a month and by e-mail on a weekly basis. Practices submitted monthly reports to coaches. Coaches spent a total of 10 months working with the clinical organizations, six months of which was spent directly working with practices.

What roles did coaches play?

The coaches served as motivators, content experts, and team facilitators. The practices were expected to take the ownership of their quality improvement initiative. Coaches acted as resources providing a broad outline of areas to address but letting the team decide sequencing and level of effort expended.

What did coaches do?

Coaches taught the CCM and Model for Improvement cycles, organized teams and team meetings, worked with leadership to reduce barriers to accomplishing the work, guided the selection of clinical measures, reviewed monthly reports, helped prioritize changes, introduced tools from the toolkit, provided examples from other settings, and acted as a resource and motivator.

Practice coaching was divided into two phases:

- Phase I: Laying the Foundation for Success (4 months)
- Phase II: Active Practice Coaching (6 months)

Phase I: Laying the Foundation for Success

The first phase of coaching took about four months and focused on laying the foundation for working with the practice teams. During this time, the coaches had three primary responsibilities:

1. INTRODUCING themselves to leadership of the organization and explaining the program and its goals, benefits, and requirements.
2. LEARNING about the organizational context of each site, including the system barriers and facilitators of quality improvement.
3. GETTING ACQUAINTED with the members of each team and generating momentum for the start of the project.

There were three major activities conducted in this phase.

ACTIVITY 1: Form Coaching Team

Your organization or initiative may have an existing group of coaches or quality improvement staff available to it, or you may be considering hiring a coach. For this quality improvement initiative, we sought coaches with experience implementing the Chronic Care Model, including population-based care using registries, self-management support, and planned care.

Coaches also should have some content knowledge about the business side of a medical practice, including operational and financial functions. We wanted individuals who would flexibly fit with a practice as well. While one person may have all these skills, we were lucky enough to have access to two coaches who together had a variety of expertise and perspectives.

ACTIVITY 2: Get Acquainted with Leadership

The coaches first contacted both the executive or middle-level leadership that initiated the quality improvement effort, as well as the local leadership ultimately responsible for implementing the work.

The primary goal of these informal conversations was for the coaches and leaders to get acquainted and discuss expectations and initial thoughts about the initiative. The following questions can be helpful conversation starters: What are you expecting to achieve during this initiative? What do you think will be the biggest barriers to success? What are you expecting to receive from us?

During these conversations, leaders were asked to provide insight into how the goals of the project would be best achieved at their site and what additional staff members should be contacted. These conversations began to develop what should be a solid and trusted working relationship between the site leadership and the coaches. The meetings also:

- Ensured that important stakeholders were brought in early, enhancing buy-in and creating the opportunity to address major problems or misconceptions early.

- Opened lines of communication directly between leaders and coaches.

- Enabled coaches to outline some of the basic requirements for successful participation, including the ability to generate population-based clinical data for monthly reports.

- Provided valuable information for coaches as they went on to develop their tactical approach; for example, when and with whom to schedule meetings for maximum attendance.

- Enabled coaches to integrate their effort with other existing system initiatives, minimizing unnecessary duplication of effort.

- Provided the executive and local leadership with enough information to be able to present the initiative to their own clinical teams. Having local leaders, rather than the coaches, motivate and introduce their teams to the effort from the very beginning sets the tone that this quality improvement work is owned by sites. The role of the coach is to support those local leaders and the teams' efforts as they move forward.

ACTIVITY 3: Orient the Practice Team to the Work

After the coaches talked with the site leadership, they introduced the effort in detail to the local practice team undertaking the quality improvement initiative. All the stakeholders who would be involved in the effort from front desk staff to physician leaders were invited to participate in this project introduction.

The more staff participating in this call, the better. For many of the practice team members, this may be the first that they have heard that they are expected to participate in a new way of working. For this reason, every effort should be made for local leadership to introduce the program. Local leaders can frame the importance of the project, provide an overview of their expectations, and offer resources to support the team.

An agenda of the phone call where the coaches and local leadership introduced the program to the practice team is in the Appendix. Note that half of the agenda is devoted to introductions and time for questions and answers. All attendees should be given a chance to participate, regardless of their position in the organization. Setting this example early can facilitate later team development.

After orienting the team undertaking the quality improvement effort, it is important for the coaches to stay in close communication with them. To build and sustain momentum, not more than three or four weeks should elapse between the time of these introductory conversations and the onsite launch of the initiative. While an effort to speak with each member of the participating practice team should be made during Phase I, do not be surprised to meet new team members during Phase II, active practice coaching. There is no substitute for an in-person orientation to get people engaged.

PHASE II: Active Practice Coaching

The second phase of the project was active practice coaching and lasted about six months. The six month design was an attempt to provide inexpensive and time-limited technical assistance to help teams get started. We know that six months of technical assistance is short compared to other quality improvement initiatives, and it may be insufficient for teams with little or no prior experience with quality improvement. This phase consisted of five activities.

ACTIVITY 1: *Introduce Prework and Prepare Practice Team for Site Visit*

As with the practice team orientation call, all members of the team and the practice leadership should participate in this meeting to introduce prework and prepare the practice team for the first site visit. In this AHRQ pilot, we conducted this call about three weeks before the learning session, allowing the site time to complete those elements of the prework that had to be done before we arrived: the clinical assessment, the financial assessment, and the Assessment of Chronic Illness Care (ACIC). Participants included the medical director of the site, administrative director of the site, physicians, nurses, medical assistants, front desk staff, and ancillary clinical staff, including dietitians and nurse care managers.

The primary purpose of this call was to discuss the plan for the upcoming site visit and to introduce the prework to the teams. However, it is likely that some new staff will participate, so it may help to conduct a brief refresher of the project and allow time for questions and answers about the general aims of the program before jumping in. Reminding the team that this is just a refresher and they can talk with other team members or leaders or e-mail questions may help keep this portion of the agenda short.

You'll notice in the companion toolkit that one of the first steps for teams when they are working to improve quality is to select measures that are important to them. Data gathered during prework is primarily for the teams' use during the learning session to decide what areas of care they first want to improve. In addition, the data provide a baseline to measure progress, an important tool for engaging senior leaders. Finally, the data provide the coaches with some insight into the needs of the teams with whom they are working. Introduce teams to the prework assessments. Examples of each of the prework assessments are available in "Integrating Chronic Care and Business Strategies in the Safety-Net" toolkit. They include:

- Clinical Assessment: Clinics start on their quality improvement journeys by selecting and measuring the outcomes for a subpopulation of patients. In the case of our initiative, the sites worked with diabetic patients, so the clinical assessment provided a baseline of clinical quality for each team's diabetic population. It is to be filled out to the extent possible through automated data. If a clinic does not have automated data, a small chart review may be necessary. Each team is expected to complete this assessment before the coaches arrive for the learning session. A copy of this assessment, called Quantitative Monthly Diabetes Report Template, is available in Key Change 2.3 in the toolkit.

- Financial Assessment: In our experience, the financial functions and performance of a practice are often fairly far removed from the daily clinical practices. In order to capitalize on possible reimbursement and cost-saving opportunities, sites can complete a financial assessment before the coaches arrive for the learning session. If multiple provider teams within one site are being coached, only one financial assessment is needed. A copy of this assessment, called Finance Collaborative Prework, is available in Key Change 2.1 in the toolkit.

- Assessment of Chronic Illness Care: This survey assesses how well teams are set up to deliver high-quality chronic illness care according to the elements of the Chronic Care Model. This survey is to be completed by each individual of the clinical team before the coaches arrive for the learning session. A copy of this survey, called Assessment of Chronic Illness Care, and a companion Scoring Guide are available in Key Change 2.1 in the toolkit.

This short prework call also provided an important opportunity to prepare the teams for what to expect during the coaches' first site visit. Be sure to allot time to discuss:

- *Completing the administrative process assessment:* This fun, poster-sized assessment assesses how well administrative processes such as answering phones and rooming patients are working. This tool is a poster-sized template that can be printed and hung on the wall. All staff and even patients are invited to place a checkmark in the box that corresponds to their perception of the processes. This assessment is completed during the coaches' first visit. A copy of this assessment, called Primary Care Practice: Know Your Processes, is available in Key Change 2.1 in the toolkit.

- *Conducting the observational assessment:* The observational assessment is designed to give coaches a sense of how the practice works with patients. During the assessment, coaches will spend a couple hours looking at the practice supports for high-quality chronic illness care: how clinical information systems and decision support are used; whether planned visits, self-management support, and linkage to community resources are conducted; and how leadership supports the team.

- *Developing the agenda for the learning session.* Before conducting this call, you should have a good sense of how you plan to structure the learning session. For more information about the learning session, see Activity 2 below. It may be helpful to share your vision and a proposed agenda for how you expect the day to go. This gives teams something to look forward to and prepare for.

- *Reaching coaches with questions.* It is likely the teams will have questions between this meeting and the first site visit about how to complete the prework, what to expect during the learning session, or other topics. Be sure to talk explicitly about how teams can reach you effectively, be it phone or e-mail.

ACTIVITY 2: Conduct the Observational Assessment (1/2 Day) + Learning Session (1/2 day)

Because the coaches did not live in the same U.S. cities as the teams they were coaching, they conducted the observational assessment and the learning session as part of the same trip. The observation assessment was conducted the afternoon of one day, and the learning session was conducted the following morning. Finding a meeting time with the team for an hour one day and then for a full morning the following was challenging. Breaking up these two functions may facilitate scheduling.

Observational Assessment

Clinical observation can be a valuable way for coaches to get a sense of how the clinic functions on an average day. In observing the flow of patients with a fresh eye, the coaches were able to identify areas where enhanced chronic illness care, such as self-management support, could be integrated with the existing operations and staffing. Using an organized observational tool helped to focus our observations in the midst of a very busy setting.

- The day began with a one-hour meeting with the team. During this meeting the coaches discussed expectations, collected prework, and administered another tool: Know Your Process (Key Change 2.1 in the toolkit).

- Coaches then observed the practices, using a standard tool to guide their observations.

- The coaches gathered the information from all assessments, including their observations, and organized it to be useful for the teams to use in setting their improvement agenda.

Learning Session

The learning session served as the big project kickoff; it was the first time the coaches met with all the teams and the site leadership face to face. The expressed purpose of the learning session was to provide an orientation to the Chronic Care Model and Model for Improvement and to help the teams get started making small-cycle changes. However, the meeting also served as a way to generate momentum for the project, and as a fun introduction to redesigning clinical care.

The coaches attempted to keep the learning session interactive, dynamic, and useful. All the baseline assessment data were presented conversationally, with coaches briefly presenting the results of the assessments and then leading the teams through a discussion about the results. Feedback sometimes got heated. Redirecting pointed questions back to the team by asking, "What do others think?" helped to diffuse energetic responses. It also set the tone that the coach is there not to fix all the practices' problems externally, but to support the team to fix their own problems. In addition, didactic presentations were kept short and substantial time was allotted for the teams to figure out how to get started doing small cycles of change. Coaches attempted to model teamwork by encouraging shy participants to speak up and share opinions. Specific content covered in the learning session is presented below.

Teaching the Chronic Care Model

The Chronic Care Model is the organizing framework around which this toolkit and coaching intervention were designed. The CCM is an evidence-based model that can help teams provide proactive, population-based care. For more on the Chronic Care Model, see the companion toolkit Key Change 1.2, Chronic Care Model Primer. Videos and PowerPoint presentations of the Model should be short, specific, and interactive. Additional examples of presentations are available at www.improvingchroniccare.org.

Reviewing Assessment of Chronic Illness Care.

By the time of the learning session, the coaches should have received all the ACIC surveys back from the team members who completed them as part of the prework. To score the ACIC, see the companion toolkit Key Change 2.1.

Assessment of Chronic Illness Care

Presenting these scores back to the group in aggregate or as blinded individual surveys gives the team members a chance to identify and discuss areas of strength and opportunities for improvement. Practices may feel discouraged when they realize how many elements of the CCM they do not currently address. Coaches familiar with quality improvement methodologies know that teams do best when they start with small changes. Reassure teams that they can make progress without addressing every element of the CCM at once. As the day progresses, teams will have a chance to discuss where they might be able to achieve early successes.

Model for Improvement

Like the Chronic Care Model, the Model for Improvement is an important organizing framework for this intervention. If the Chronic Care Model is what the teams are going to work on, then the Model for Improvement is how the teams are going to do the work. Plan-Do-Study-Act cycles are the key component of the Model for Improvement, and there are many creative ways to present this content, including games. For more information on the Model for Improvement, see the companion toolkit Key Change 1.2, A Model for Accelerating Improvement. Don't be concerned if not everyone "gets it" all at once. This is just an introduction; these concepts are best learned by doing.

Observational Assessment Results and Group Discussion

During this time, coaches present qualitative feedback to the teams about what they observed during their observational assessment. A good approach is providing an overview of what you observed the teams doing well and then identifying areas where easy enhancements could be made to better address patient needs. For example, if patients are routed through some sort of nurse- or medical assistant-led checkout process before leaving the office, perhaps goal setting or action planning could be integrated. This exercise is most helpful when coaches can point out potential solutions simultaneously with potential areas for improvement.

Where to Start

After learning about the concepts behind the Chronic Care Model, teams often wonder how to get started. Here, the coaches introduced a menu of starter ideas, areas that the team might like to address first. This was not a prescriptive list, but it was meant to start discussion. This was the most valuable and important aspect of the learning session: the time teams had together to brainstorm Plan-Do-Study-Act (PDSA) cycles and how they would make the program run.

The Toolkit

The companion toolkit provides a sequenced approach to help teams improve care. It also provides content and tools for almost any related topic of interest from selecting a registry to trying out planned visits. During this session, coaches provided an interactive overview of the toolkit with a special emphasis on its approach to the business case for improved care. The toolkit is available to teams free of charge online at both www.improvingchroniccare.org and www.ahrq.gov. Team members can use any of the tools or review content on the Web without printing out a heavy binder.

Monthly Reporting

Coaches also briefly introduced the monthly reports teams were required to submit to them. Key Change 2.3 of the Integrating Chronic Care and Business Strategies in the Safety Net toolkit provides examples of the Quantitative Monthly Diabetes Report Template and the Narrative Monthly Report Template that the teams completed. These monthly reports serve several functions. They provide a tangible deliverable and an opportunity for the teams to ask questions of the coaches in a systematic way. The reports also provide a template for the teams to look at changes in health process and outcome measures as a result of their work. They also demonstrate evidence of improvement to be used to engage leadership or other teams in spread.

Planning Future Team Meetings

In order for teams to successfully make changes in how they deliver care, regular time needs to be set aside for the team to gather together. This can take the shape of a weekly one-hour meeting or a series of short, daily huddles. Either way, it is important to establish a time to share what has been learned, develop new ideas to test, and maintain momentum. Because trying to get started on a new initiative in the midst of a very busy clinical schedule can be challenging, the time set aside by the coaches must be more than just didactic presentations. It must be value-added planning time for the teams as well.

Evaluation

In the spirit of continuous quality improvement, the coaches asked the teams to evaluate them after the learning session. The evaluation form, "Tell Us What You Think," can be found in the Appendix.

ACTIVITY 3: *Coaching Through Regular Team Meetings*

After the learning session, teams start trying to improve care using PDSA cycles. The coaches participated by phone in the team's regular weekly or biweekly meetings, though in-person participation could also work. The expectation was that a team leader would facilitate the meetings, but the coaches were available before the meeting to brainstorm an agenda, during the meeting to provide suggestions and ideas, and after the meeting to reflect on how to best move the project along. The team leader can be anyone on the team who is able and interested in convening team meetings, maintaining momentum for the initiative, and overseeing the implementation of change ideas. Some teams have one person who acts as the team leader, such as the medical director of the practice or the office manager. Other groups rotate team leadership among team members. For more information about leadership, see Key Change 1.1 in the toolkit, Organize your lead quality improvement team.

Coaching through these regular meetings, as opposed to establishing separate meetings either individually or as a group, has many advantages. First, since the teams are already meeting, coach participation is efficient. If the teams have questions, especially at the beginning as they work on PDSAs, they can get coaching help and ideas right way. Participating in team meetings also enables the coaches to see how the project is progressing. If, for example, key members of the team are not attending, the coaches can talk with leaders who may be able to encourage attendance.

Initially, the coaches provided substantial guidance, but over time, the meetings shifted to be led and managed much more independently. From the beginning, an important goal was for the teams to own the meetings and to perceive the coaches as a support but not an active "implementer" or team member. Coaches do not and cannot know the local politics and organizational context as well as the team members do, and they are only available to the teams for a limited time.

The coaches also provided ad hoc support to individual members of the teams through e-mails and phone calls. Often this involved providing a link to a specific tool or a recommendation for a speaker or training on a topic of interest. Sometimes, the coach acted as a listening ear when people felt frustrated or unable to move forward. The coaches took on various roles throughout the six-month active coaching phase of the project: observer, trainer, meeting participant, report-reviewer, and ad hoc resource. These roles changed as the needs of the teams changed.

Having clear, well-communicated boundaries about what is and is not the job of the coach is important. Coaches should:

- Be in a position of offering ideas, not imposing what they want to get done.
- Help the teams actually implement what they learn.
- Set up systems for the benefit of the clinic and its staff, not the organizing group, the coaches, or even the leadership.

Finally, there is only so much a coach can do. To be successful, coaching has to be sufficiently supported and matched by good leadership, sufficient resources, and a clear idea of the desired outcomes. Some organizations and teams are just not ready or able to make good use of a coaching resource.

ACTIVITY 4: Communicating with Leadership

In addition to participating in team meetings via phone, or occasionally in person, coaches also worked with local and executive leadership, communicating about the project, highlighting challenges and successes, and helping leaders think about how they could contribute to the success of the effort. Sometimes this meant drawing attention to resistant staff or broken systems that impeded the ability of the team to move forward; other times it was encouraging leaders to ask and follow up with the teams about their work.

ACTIVITY 5: Closing out Coaching

Our intervention was deliberately low-intensity and lasted six months, though certainly you could continue coaching if interest or funding were available. In preparation for the last team call, teams were asked to discuss two things: first, reflect on the initiative and how it affected their relationships both with patients and coworkers, and two, think through how the effort would be sustained after the coaching component ended. During the meeting, the teams presented on these topics and the coaches reminded the teams about available resources, including the companion toolkit.

Suggested Modifications to our Practice Coaching Approach

The aforementioned coaching intervention was evaluated as part of AHRQ's "Integrating Chronic Care and Business Strategies in the Safety Net" project. RAND assessed the implementation of the intervention through site visits to the two participating medical centers and interviews with key informants. Below we offer the following suggestions for modifying the practice coaching intervention.

1. *Coaching should include more face-to-face interactions.* Due to the ease of communication and discussion, the pilot site participants believed that

they would benefit from more frequent in-person contact with the coaches. Although for the most part the telephone calls and e-mail functioned well, some participants felt their enthusiasm was dampened when the coaches could not be reached.

2. *An internal coach might be added.* The participants felt that sometimes the external coaches' advice was too general and not applicable to their particular organizational setting. In one site, a physician who had prior experience using the Chronic Care Model was consulted by others about how to implement specific changes. Hence, many participants suggested that an internal coach who knows their system better and is more readily available could complement an external coach. It was also noted that an internal coach should be given sufficient time and clear responsibility, so as not to cause antipathy among other staff members.

3. *Coaching intensity may need to be greater at the beginning.* The meeting and coaching time allotted was perceived to be insufficient for participants to learn, ask questions, and exchange information. The participants commented that they needed more help at the beginning and suggested greater intensity of coaching until they became self-sufficient. It was also suggested that everyone in the practice who plays a role in CCM implementation should be invited to the first in-person coaching meeting. Some recommended that the coaches provide a more specific timeline for changes.

4. *Coaches should be more proactive and creative in introducing the toolkit.* The interviewees suggested that coaches be more proactive in introducing the toolkit. The learning session could allot more time to reviewing the toolkit to increase users' understanding of its contents. One participant suggested that the coaches could create scenarios to demonstrate how and when to use the toolkit. Others suggested that the coaches remind them to use the toolkit.

5. *Continue coaching for a longer period of time.* We designed the coaching intervention to get the practice team started in CCM implementation, but the coaching was perceived to be worth continuing beyond the six-month timeframe.

Acknowledgements

The authors gratefully acknowledge Ed Wagner, Brian Austin, and Dona Cutsogeorge from the MacColl Institute for Healthcare Innovation; Wendy Jameson and Hunter Gatewood from the California Health Care Safety Net Institute; and Anne Tillery and Sarah Bylsma at Pyramid Communications who helped to develop, edit, and format this work. The authors also thank all of the many individuals who contributed by sharing their expertise and reflections on practice coaching, including:

Mike Hindmarsh, MA Consultant, MacColl Institute for Healthcare Innovation, Principal, Hindsight Healthcare Strategies

Rick MacCornack, PhD Chief Systems Integration Officer, Northwest Physicians Network Director, South Sound Health Communication Network

Terry McGeeney, MD, MBA President and CEO, TransforMED

Marjorie M. Godfrey, PhD(c), MS, RN Instructor, The Dartmouth Institute for Health Policy and Clinical Practice, Director, The Clinical Microsystem Resource Group

Nicole Van Borkulo, Med Principal, NVB Consulting Inc., Consultant, Washington State Collaborative for Better Health

Darren A. DeWalt, MD, MPH Assistant Professor of Medicine, University of North Carolina at Chapel Hill, Quality Improvement Consultant, Improving Performance in Practice

Peter Margolis, MD, PhD Professor of Pediatrics, Cincinnati Children's Hospital Medical Center, Center for Health Care Quality, Quality Improvement Director, Improving Performance in Practice

Patricia L. Bricker, MBA State Director, Pennsylvania Improving Performance in Practice

Allyson Gottsman Associate Director and Improving Performance In Practice Director, Colorado Clinical Guidelines Collaborative

Marjie Harbrecht, MD Medical/Executive Director, Colorado Clinical Guidelines Collaborative

Julie Schilz, BSN, MBA Manager, Improving Performance in Practice and Patient Centered Medical Home, Colorado Clinical Guidelines Collaborative

Zula Solomon, MBA Improving Performance in Practice and Patient Centered Medical Home, Quality Improvement Coach, Colorado Clinical Guidelines Collaborative

Linda Lawrence Cade, MSN, NP, CDE Practice Redesign Coach, Humboldt-Del Norte, Independent Practice Association and Open Door Community Health Centers

Alan Glaseroff, MD Chief Medical Officer, Humboldt-Del Norte Foundation

Mary Ruhe, RN, MPH EPOCHS Project Coordinator, Family Medicine Research Division at Case Western Reserve University

Kurt C. Stange, MD, PhD Professor of Family Medicine, Epidemiology and Biostatistics, Oncology and Sociology, Case Western Reserve University

References

1. Asch SM, Baker DW, Keesey JW, et al. Does the collaborative model improve care for chronic heart failure? Med Care 2005;43(7):667–75.

2. Baker DW, Asch SM, Keesey JW, et al. Differences in education, knowledge, self-management activities, and health outcomes for patients with heart failure cared for under the chronic disease model: the improving chronic illness care evaluation. J Card Fail 2005;11(6):405–13.

3. Chin MH, Drum ML, Guillen M, et al. Improving and sustaining diabetes care in community health centers with the Health Disparities Collaboratives. Med Care 2007;35(12):1123–5.

4. Chin MH, Cook S, Drum ML, et al. Improving diabetes care in Midwest community health centers with the Health Disparities Collaborative. Diabetes Care 2004;27(1):2–8.

5. Pearson ML, Wu S, Schaefer J, et al. Assessing the implementation of the chronic care model in quality improvement collaboratives. Health Serv Res 2005;40(4):978–96.

6. Mangione-Smith R, Schonlau M, Chan KS, et al. Measuring the effectiveness of a collaborative for quality improvement in pediatric asthma care: does implementing the chronic care model improve processes and outcomes of care? Ambul Pediatr 2005;5(2):75–82.

7. Schonlau M, Mangione-Smith R, Chan KS, et al. Evaluation of a quality improvement collaborative in asthma care: does it improve processes and outcomes of care? Ann Fam Med 2005;3(3):200–8.

8. Vargas RB, Mangione CM, Asch S, et al. Can a chronic care model collaborative reduce heart disease risk in patients with diabetes? J Gen Intern Med 2007;22(2):215–22.

9. Reich Y, Ullmann G, van der Loos M, et al. Coaching product development teams: toward a conceptual foundation. RED 2008; in press.

10. Nagykaldi Z, Mold JW, Robinson A, et al. Practice facilitators and practice-based research networks. J Am Board Fam Med 2006;19:506–10.

11. Gandhi TK, Puopolo AL, Dasse P, et al. Obstacles to collaborative quality improvement: the case of ambulatory general medical care. Int J Qual Health Care 2000;12(2):115–23.

12. Green PL, Plsek PE. Coaching and leadership for the diffusion of innovation in health care: a different type of multiorganization improvement collaborative. Jt Comm J Qual Improve 2002;28(2):55–71.

13. Margolis PA, Lannon CM, Stuart JM, et al. Practice based education to improve delivery systems for prevention in primary care: randomized trial. BMJ 2004;328(7436):388.

14. Rosenthal MS, Lannon CM, Stuart JM, et al. A randomized trial of practice-based education to improve delivery systems for anticipatory guidance. Arch Pediatr Adolesc Med 2005;159:456–63.

15. Nagykaldi Z, Mold JW, Aspy CB. Practice facilitators: a review of the literature. Fam Med 2005;37(8):581–8.

16. Goodwin MA, Zyzanski SJ, Zronek S, et al. A clinical trial of tailored office systems for preventive service delivery: the Study to Enhance Prevention by Understanding Practice (STEP-UP). Am J Prev Med 2001;21(1):20–28.

17. Cohen D, McDaniel RR, Crabtree BF, et al. A practice change model for quality improvement in primary care practice. J Healthc Manag 2004;49(3):155–68.

18. Deyo RA, Schall M, Berwick DM, et al. Continuous quality improvement for patients with back pain. J Gen Intern Med 2000;15:647–55.

19. Yano EM, Rubenstein LV, Farmer MM, et al. Targeting primary care referrals to smoking cessation clinics does not improve quit rates: implementing evidence-based interventions into practice. Health Serv Res June 3 [Epub ahead of print] 2008; in press.

20. Rycroft-Malone J, Harvey G, Seers K, et al. An exploration of the factors that influence the implementation of evidence into practice. J Clin Nurs 2004;13:913–24.

21. Stange KC, Goodwin MA, Zyzanski SJ, et al. Sustainability of a practice-individualized preventive service delivery intervention. Am J Prev Med 2003;25(4):296–300.

22. Sullivan G, Duan N, Mukherjee S, et al. The role of services researchers in facilitating intervention research. Psychiatr Serv 2005;56:537–42.

23. Hackman JR, Wageman R. A theory of team coaching. Acad Manag Rev 2005;30(2):269–87.

24. Stetler CB, Legro MW, Rycroft-Malone J, et al. Role of "external facilitation" in implementation of research findings: a qualitative evaluation of facilitation experiences in the Veterans Health Administration. Implement Sci 2006;1:23.

25. Hagedorn H, Hogan M, Smith JL, et al. Lessons learned about implementing research evidence into clinical practice: experiences from VA QUERI. J Gen Intern Med 2006;21(suppl 2):S21-4.

26. Sales A, Smith J, Curran G, et al. Models, strategies, and tools: theory in implementing evidence-based findings into health care practice. J Gen Intern Med 2006;21(suppl 2):S43-9.

27. Curran GM, Mukherjee S, Allee E, et al. A process for developing an implementation: QUERI Series. Implement Sci 2008;3:17.

28. Litaker D, Tomolo A, Liberatore V, et al. Using complexity theory to build interventions that improve health care delivery in primary care. J Gen Intern Med 2006;21:S30-41.

29. Miller WL, McDaniel RR, Crabtree BF, et al. Practice jazz: understanding variation in family practices using complexity science. J Fam Pract 2001;50:872-8.

30. Ruhe MC, Weyer SM, Zronek S, et al. Facilitating practice change: lessons from the STEP-UP clinical trial. Prev Med 2005;40:729-34.

31. Stroebel CK, McDaniel RR, Crabtree BF, et al. How complexity science can inform a reflective process for improvement in primary care practices. Jt Comm J Qual Patient Saf 2005;31(8):438-46.

32. Agency for Healthcare Research and Quality. Practice enhancement assistants improve quality of care in primary care practices. Available at: http://www.innovations.ahrq.gov/content.aspx?id=1768.

33. Margolis PA, McLearn KT, Earls MF, et al. Assisting primary care practices in using office systems to promote early childhood development. Ambul Pediatr 2008;8(6):383-7.

34. Hogg W, Baskerville N, Lemelin J. Cost savings associated with improving appropriate and reducing inappropriate preventive care: cost-consequences analysis. BMC Health Serv Res 2005;5(1):20.

35. Baskerville NB, Hogg W, Lemelin J. Process evaluation of a tailored multifaceted approach to changing family physician practice patterns and improving preventive care. J Fam Pract 2001;50(3):W242-9.

36. Lemelin J, Hogg W, Baskerville N. Evidence to action: a tailored multifaceted approach to changing family physician practice patterns and improving preventive care. CMAJ 2001;164(6):757-63.

Customer Complaints as a Source of Customer-Focused Process Improvement: A Constructive Case Study

Kari Uusitalo, Henri Hakala and Teemu Kautonen

ABSTRACT

Process-based thinking commonly focuses on enhancing the efficiency of processes, while it is often criticized for not paying enough attention to the customer. This paper argues that customer complaint information can be used as a basis for customer-focused process improvement. Thus, it is not enough to make the complaining customer satisfied, but the complaint information should also feed back to the actual processes where the fault causing the complaint arose and where it can be removed. The empirical component of the study includes the development of a novel construction to utilize customer complaints for process improvements, which was implemented in a large

Finnish enterprise operating in the wholesale logistics environment. The results show benefits at both operational and strategic levels.

Keywords: customer orientation, process improvement, customer complaints, complaint management, operations management, constructive method

Introduction

The market-oriented philosophy in marketing and management literature has emphasized customer satisfaction and loyalty as sources of performance and profitability (e.g. Deshpandé, Farley and Webster, 1993; Foster, Gupta and Sjoblom, 1996; Jaworski and Kohli, 1993; Knox, 1998; Oakland and Oakland, 1998; Slater and Narver, 1996). However, the customer orientation seldom reaches the operational level of business processes in theory or practice. Process-oriented management teachings such as activity based management (Turney, 1992), total quality management (Creech, 1994; Mizuno, 1992), business process re-engineering (Earl and Khan, 1994), continuous improvement (Davenport, 1993), lean management (Taylor, 1999; Vollmann, Berry and Whybark, 1997) and supply chain management (Shapiro and Heskett, 1985) have traditionally focused on enhancing the efficiency of processes within organizations. While a number of scholars have raised the role of the customer in the improvement of business processes (Jones and Sasser, 1995, Kohli and Jaworski, 1990, Slater and Narver, 1994, 1996), these teachings have also been criticized for being rhetoric and not paying enough genuine attention to the customer (e.g. Wood, 1997). In short, despite their development towards increased customer focus, these 'engineering' approaches essentially concentrate on processes as such and do not appear to provide sufficient support to focus on the issues that are important to the customer.

This paper argues for a customer-focused approach to the improvement of business processes by developing a construction which systematically utilizes customer feedback in form of complaints to achieve process improvements both at strategic and operational levels. The basic idea is that it is not enough to make the complaining customer satisfied, but that the complaint information should feed back to the actual processes where the fault causing the complaint arose and where it can be removed, thus avoiding further similar errors. This thinking is essentially utilizing the ideas of a learning system (Checkland, 2000) and feedback loops that balance the variety between the environment and the operations (Beer, 1985). While complaint management has been addressed in the previous literature (e.g. Boshoff, 1997, 1998; Brown, Cowles and Tuten, 1996; Feinberg, Widdows, Hirsch-Wyncott and Trappey, 1990; Hart, Heskett and Sasser, 1990;

Johnston, 1995), Johnston and Mehra (2002) emphasize that further research is required especially with respect to the 'how', that is, how the complaint information could be utilized operationally. Moreover, little research has been devoted so far to investigating how companies can better utilize qualitative customer complaint information. The present paper addresses both of these knowledge gaps.

Empirically, the paper adopts an applied approach based on the constructive case study method. The constructive method concentrates on developing and implementing a new, innovative and theoretically anchored construction (e.g. a model, plan, organization, technology, software or a combination of these) to solve a real-world problem situation (Kasanen, Lukka and Siitonen, 1993; Lukka, 2000, 2003, 2005). The implementation phase is an integral part of this method, as the ideal construction not only makes a theoretical contribution but also solves the practical problem (Lukka, 2000). Thus, in business studies the construction is subjected to the practical test of whether it works in the company or not. The construction developed in the present study includes a database solution for collecting and analysing qualitative customer complaint data in a large Finnish company operating in the wholesale logistics environment. The aim of the construction was to provide a tool for customer-focused process improvement.

The contribution of this paper is two-fold. Firstly, it introduces a novel construction which links customer complaints to the company's processes arguing that complaint information can be effectively used to improve customer focus and operational quality. Secondly, from a managerial point of view, the paper describes a construction that effectively utilizes customer complaint information in support of managerial decision making both at operational and strategic levels aiming towards improved operational quality.

The paper is arranged following the logic of the constructive method. The first section reviews literature relevant to analysing the role of customer complaints as a source of information for the purpose of process improvement. This review summarizes the main literature used in developing the construction. Next, the constructive case study methodology is described and the case company and the developed construction are introduced. Finally, the results of the study are presented and discussed, followed by the conclusions and implications for management and further research.

Customer Complaints and Process Improvement

Extending the market-oriented philosophy to the management of processes would imply that the emphasis should be placed on identifying and improving those processes in the company's value chain that generate the most value to the customer. Such improvements are not just a cost but also an investment in long-term profitable customer relationships (Reichheld and Sasser, 1990). However, errors and unsatisfactory service occur in all businesses given that "mistakes are an unavoidable feature of all human endeavour" (Boshoff, 1997, p. 110). Occasional failures are not necessarily bad. As a matter of fact, most customers accept that things go wrong sometimes and are happy enough as long as the problems are solved and do not occur again (Bitner, Booms and Tetreault, 1990; Feinberg et al., 1990).

Different types of faults can be prioritized on the basis of the cost they cause to the company or its customers (Albright and Roth, 1994; Shank and Govindarajan, 1994). The Japanese quality philosophy distinguishes between random and systematic faults in this context (Mizuno, 1992). Random errors often have relatively simple causes and are thus fairly easy to identify and analyse. They are often 'human' and can therefore usually be corrected by the person responsible for the particular task (Cardy and Dobbins, 1996; EFQM, 1997; Oakland and Oakland, 1998). Systematic errors cause the customer to experience dissatisfaction on a continuous rather than sporadic basis. The reasons causing this type of errors are often multifaceted and removing their causes requires complex analysis (Mizuno, 1992). To correct the problem is therefore a task for the management who have the power to eliminate the causes. Our particular interest in this paper concerns systematic errors as removing such faults in a company's processes demonstrates the greatest potential to improve quality in a way valued by the customer (Berry and Parasuraman, 1997; Clinton and Hsu, 1997; Hammer, 1990).

The identification of systematic faults requires a considerable amount of versatile data (Reichheld and Sasser, 1990). We argue that customer complaints can be a valuable and inexpensive source of information for identifying systematic errors and enabling customer-focused process improvement. The feedback that the customer provides out of their own initiative, such as complaints, is often very direct, concrete and detailed (Reichheld and Sasser, 1990). Thus, compared with data collected by means of customer surveys and panel studies, complaint information provides a more reliable picture of the customer's true opinion. Complaint information has several managerial applications. Johnston (2001) developed and tested a conceptual model demonstrating three routes that link complaint processes with the company's financial performance. We coined these routes as the Customer Orientation, the Human Resource and the Engineering routes (Figure 1).

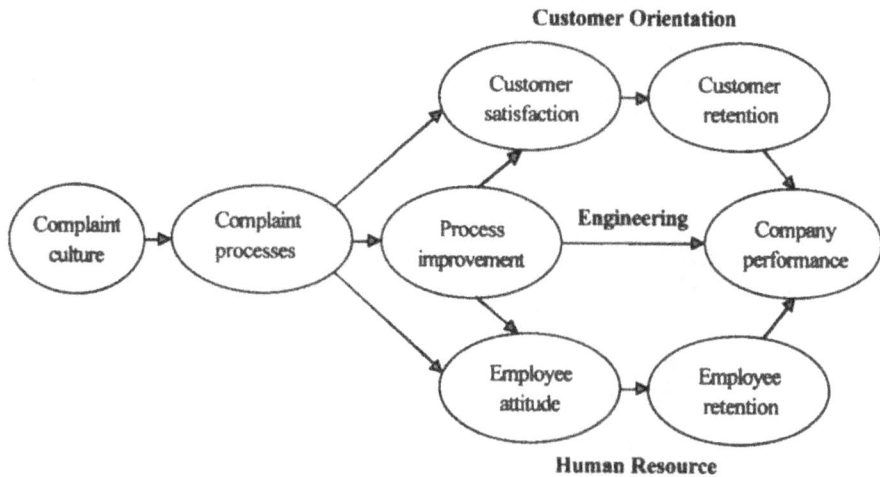

Figure 1: Three routes linking complaints to performance (based on Johnston 2001)

The Customer Orientation route suggests that complaint processes impact on customer satisfaction, which in turn through its effect on customer retention has an impact on the firm's financial performance. It essentially covers the customer recovery process, that is, making the complaining customer satisfied through an appropriate correction of the error made. Furthermore, Johnston (2001) argued that complaint processes, if made 'staff-friendly,' can also have an impact on employee attitudes and to financial performance via employee retention. We call this the Human Resource route. The reasoning behind this route suggests that by making complaint management easier to the employees, allowing a certain degree of human error, and relaying not only the complaints but also the positive feedback received from the customers to the employees, employees are believed to be happier, learn from their mistakes and remain with the company, thus reducing operation and switching costs. Finally, Johnston (2001) suggested that complaint processes should be designed to focus on process improvements that are likely to achieve savings and thus positively impact profitability, which is not necessarily the case if the improvement merely targets customer satisfaction. These elements form the third route in the model, which we coined the Engineering route. According to our understanding, this is the process where the errors causing the complaints are identified, analysed and tracked back to their source. Thereafter the information can be used in aid of decision making in an attempt to improve processes, thus preventing similar errors happening again.

The focus in the following analysis is on the Engineering route. We argue, however, that the three routes are intertwined. Process improvements affect customer satisfaction and retention, as they do employee attitudes and retention.

Similarly, process improvement should be based on information about the factors that have a positive impact on customer satisfaction and employee attitudes. The particular emphasis of the following analysis is on utilizing customer complaint information to determine what makes the customer dissatisfied and using this as a basis for process improvement, which in turn aims to avoid the repetition of the errors that gave rise to the complaint in the first place. So far little research has been aimed at finding tangible methods to analyse and derive operational benefits from customer complaints. The following section presents a construction that was implemented in a large Finnish company as a solution for analysing complaint information and tracking back complaints to the processes within the organization where the fault causing the complaint occurred. This is followed by a discussion of the results achieved through the implementation at both the operational and strategic levels.

Method and Data

The Constructive Case Study Method

The constructive method is a specialized form of case study, which concentrates on developing a new, innovative construction to solve practical, real-life problems Lukka, 2003, 2005). Lukka (2000) characterizes construction as an abstract concept which has a nearly infinite number of possible realizations. Examples include different models, diagrams, plans, organizational structures, commercial products and information systems. A particular characteristic of constructions is that they are not discovered but invented. The construction developed in the present study is a novel, practically relevant method of systemizing the utilization of customer complaint information for processes improvement.

In addition to building a theoretically anchored construction, the implementation of the developed construction is at the core of the method. Therefore, unlike for example in action research, the researcher does not attempt to be an observing bystander but works actively and explicitly on the project in order to make it work in practice (Lukka, 2005). A constructive study is experimental by its nature. Following the pragmatic philosophy of science, the constructive method believes that one can make a contribution to theory through a profound analysis of what works and what does not work in practice (Lukka, 2000). The ideal results from a constructive case study combine both the solution to the practical problem and a contribution to theory. The theoretical contribution can take the form of an entirely new theory but more often constructive studies demonstrate, test or develop existing theory (Keating, 1995; Lukka, 2000, 2005).

The present study was carried out in three phases. The pre-study phase and the development of the construction took place mostly during 1999-2001, whilst the implementation of the construction in the case company followed in 2001-2002. Analysis and monitoring of the continuing implementation of the construction has continued ever since and the construction is still in use at the case company at the moment of writing this paper (August 2007). The relatively long period between the initial implementation in the case company and this analysis, strengthens the validity the construction and findings, as the construction has been not only initially successful, but stood against the test of time in ever changing world. The strong intervention required by the constructive case study method was enabled by the main project researcher's position as a project manager responsible for the development of the customer feedback system, and his previous experience with the company. The credible organizational role of the main researcher allowed him to collect a wealth of data by means of observation, participation and interaction with various staff members in meetings, informal discussions and email dialogues. The researcher made extensive notes over the course of the implementation process, which were used as a basis for the present analysis. The following sections present the case company and the theoretical construct, followed by a presentation and discussion of the results of the study.

The Case

The case company 'HouseTech Corp' (pseudonym) is one of the major agents in the technical wholesale in Northern Europe. It generates an annual turnover of around one billion euros, has approximately 2500 employees and maintains operations in eight countries in Northern Europe. As is typical in the wholesale business, logistics has a central role in HouseTech Corp's operations. Its role in the value chain is to act as an intermediary between the manufacturers and their customers. The present study was conducted in HouseTech Corp's central distribution centre in Finland. This unit maintains a product range of nearly 30 000 items and delivers approximately six million order lines per annum. Its clientele includes electrical, heating, plumbing, ventilation, air conditioning and refrigeration contractors, industrial companies, power plants, public organizations and retailers. The central distribution centre delivers goods with the help of transport partners directly to business or public sector customers or the customers can choose to collect the goods from the central warehouse. Moreover, the company operates a chain of 'express stores' where local small businesses can purchase and collect a limited selection of items without prior order.

This study deals with the customer delivery process. The customer delivery process is a combination of the physical movement of goods and information.

Quality in this operation means, from the customer point of view, that the ordered goods reach the right place undamaged and at the right time. Due to the nature of the technical wholesale business, the majority of complaints relate to actual 'physical' shortcomings in the delivery process rather than emotional perceptions of service quality. In other words, the complaints relate to situations where the products ordered have reached the customer late, in a faulty condition, in insufficient quantities, or the correct products have not reached the customer at all. These sorts of problems are common in all high volume warehousing operations. Most customers tend to complain about this kind of shortcomings because the operations management system automatically invoices all the goods that have left the warehouse. HouseTech Corp had and still has a back office function to handle complaints that cannot be immediately and informally solved by the sales clerk. While sales clerks input most of the complaints into the system, further processing and the correction of errors is dealt with by the back office complaint handlers.

The need for the project on which this paper is based arose because the logistics management was concerned about the increasing number of complaints and the associated costs. The preliminary analysis of the logistics management concluded that the quality management projects already undertaken—including a quality system covering all functions in accordance with the ISO9000 standard and self-assessments in accordance with the criteria provided by the European Foundation for Quality Management—did not seem to focus on the issues that the customers complained about. At the operational level, the complaint handlers were not satisfied with the IT system and had concerns over the quality of the complaints process. The complaint handlers are operational problem-solvers whose task is to solve the problem for the customer, make the appropriate corrections in the warehousing and invoicing systems as well as find out who is responsible for the direct cost associated with the error, whenever this is humanly possible. Complaints handlers did not have a managerial perspective to their work, but a lot of tacit knowledge on what caused the problems in the first place. Since the complaint data was neither systematically collected nor analysed and the complaint handlers merely focused on correcting individual delivery errors, the complaint information and the complaint handlers' tacit knowledge did not reach the managerial level. An effective method and respective working procedures were required to address this problem.

The logistics director of HouseTech Corp had a strong belief in the benefits of using customer data, but no clear vision as to how one should go about realizing the benefits. At the time, there was no known off-the-shelf software or other solution available for the problem situation at hand. This is why the company decided to have a research-based solution developed to solve the problem. The aim was to provide a solution to the question of how to use the complaints information in order to improve quality in the processes causing the complaints in the first place. As the precise objectives of the project could not be clearly defined, the project was given plenty of room for innovation. The research process could be characterized as heuristic. As the objectives were not clear, the best means of getting there could not be 'programmed' and planned in advance (e.g. Moustakis, 1990). Moreover, numerous small problems had to be solved during the course of the project as they emerged, and many of these solutions needed to be approved by a number of people in the organization, which slowed down the process. Against this backdrop, it was necessary to choose a method where the problems, disturbances and unexpected difficulties could be tackled heuristically as they appear (Wisner and Kuorinka, 1988), and the constructive methodology appeared appropriate and flexible enough for this purpose.

The Construction

To function as the technical core of the construction, a database solution was created based on standard relational databases and the SQL protocol. The user interfaces were custom-developed, and although the system was built on standard database elements, the application itself was new and not available off-the-shelf at the time. The user interface for the input of data into the system was created around the job description of the complaint handlers. This was done because the complaint handlers were the ones holding the most information in each individual case and it was recognized that it would be beneficial to capture some of their knowledge to support managerial decision making. In the new system, a qualitative description of what went wrong from both the customer's and the sales clerk's perspective was made available to the complaint handler. A new task for the complaint handler became to link the complaint to the company's process and activity descriptions in the database system. Furthermore, since complaint handlers obviously find out what happened and why as well as how the complaint was solved, the new system required them to record their own description of these elements too. All of the complaint related details, although recorded mainly for the needs of the complaint handler, were stored in the data warehouse. All the details, including the qualitative customer comments, could be easily accessed should someone wish to look deeper into the individual complaints later on.

The following information was put together for each customer feedback event:

- Complaint information from the customer, what has happened and the customer's perception as to why this has happened.

- Sales clerk's immediate reaction (e.g. calling the customer to find out what happened exactly), possible corrective action and interpretation of the event.
- Order data.
- Complaint handler's description and analysis of the complaint's causes and effects, and a description of the corrective actions taken.
- Link to activities and agents identified to have been part of the cause for the complaint.

The system recorded both negative and positive customer feedback in a similar manner. As much as 17 % of the feedback was positive. These customers could be described as particularly delighted about the service, given that they made the effort to specifically relay their satisfaction back to the organization. The rest of the feedback consisted of complaints, which can be classified roughly into errors caused by workers or local poor working methods (40 %), poor internal processes (10 %) and poor external processes (50 %, e.g. deliveries by the transport partners).

However, the real novelty of the construction does not lie in its technical structure, user interfaces or what data is recorded, but in its managerial aspects. Three contributions of the construction should be particularly emphasized in this context. First, the new system classifies the errors that have caused the customer complaint for the purposes of further analysis and aid in managerial decision making. Second, the system allows the management to trace back the complaint and the respective error to the procedures, individuals, vehicles, partners (transport companies) or machines responsible for the error within the whole delivery process. Thus, with the help of the new system, the management can aim process improvement actions to those processes that really matter to the customer. Third, the new system allows the company to combine the complaint data with other information already available in the company. The procedures and activities that the complaints are now linked to were already accurately defined in the company and used by its activity-based cost accounting system. As the whole company utilizes the same data warehouse, the construction now also links complaints directly with cost accounting, and provides qualitative data for the uses of management accounting. The following section discusses the impact of the construction at operational and strategic levels as it was implemented in HouseTech Corp.

Results

Operational Level – Complaint Handlers

The implementation of the construction had several positive effects at the operational level of complaint handling. As an immediate effect of utilizing the new construction, the complaint handlers perceived the recording, analysing and reporting of complaints data to have become easier, making their work easier and more motivating.

"The best thing in the new system is that customer feedback, with all associated order data, prints out automatically in our office. Just by reading one paper or screenful, I can now get a whole picture of the complaint, and start doing my job without the need to search for more information from the computer systems or telephone around the company and bother other people with simple questions. The system makes our work considerably easier and faster. Another great thing in the new system is the feeling one gets while recording one's own actions into the database, that the work we do is not wasted—someone is going to use the information later on. Feels like the bosses have finally realized how important work we are doing!"(A complaint handler, April 2002)

Therefore, in terms of the conceptual model (Figure 1), the process improvement in fact feeds back to the employee attitudes component in the Human Resource route which, the model predicts, contributes to performance via employee retention.

Moreover, the average working time used for handling each complaint was measured to have reduced by approximately 15 % (or eight minutes) due to the new construction. The measurement was conducted by recording the working time used on the different phases of the customer complaint process during one week each in June 1998 and November 2004 by means of work-time clocking and time-logs recorded by the new IT system itself. This result is a direct measure of success in terms of improving the effectiveness and efficiency in the process of complaint handling. Although no direct cost savings were made through staff redundancy, the direct time saving, for its part, made it possible to keep staff numbers constant despite growth in operations volume.

Operational Level – Warehouse Management

The warehouse management (the lowest level of operational management) found the information created by the construction a very useful tool. With the implementation of the construction, these managers started to review the complaint data systematically and continuously, whereas previously their knowledge about complaints was based on sporadic discussions with the complaint handlers. They could now better monitor the performance in terms of the number of complaints tracked back to their area of responsibility. The following quote from one of the warehouse operations managers illustrates this:

"We have had the custom of going through the errors with employees every week. Occasionally, there were unpleasant situations where complaint handlers had marked an error as caused by a particular employee, but the employee himself denied responsibility. And who would want to take the responsibility for errors as they are linked to the productivity bonus and therefore to the worker's pay check. We as supervisors must be able to prove the error and its link to the specific employee in a reliable manner if required, but previously without the system it was quite difficult. The new system enables us to print out every complaint with the associated data that shows extensively the cause for the error, who made it and what sort of hassle the error caused. An extensive and well documented description gives less room for guesswork and speculation, a thing that my workers have started to appreciate. Our work becomes much easier; we can take action backed by facts instead of guessing and shooting from the hip." (A warehouse operations manager, August 2002)

Furthermore, the new system enabled warehouse managers to effectively analyse the reasons behind each complaint and combine this information with their own detailed operational knowledge. Thus, the system provided warehouse management with factual information backing their decisions to change working methods, relocate employees or initiate training. For example, when an employee was identified as the cause for a systematic fault, the warehouse management initiated discussions with the respective employee regarding the problem. In some cases this was enough, if the fault was caused for instance by unintended carelessness on part of the employee. In others, the reason could be traced back to faulty equipment, a local working method or a plain misunderstanding. Where the underlying cause was identified as shortage of skill, the warehouse manager could rotate the employee to other duties more suitable to their skills or initiate further training.

Moreover, as positive feedback was recorded in a similar manner as complaints, the warehouse management adopted the habit of reviewing also this information and giving feedback to employees on their successful efforts that had led to customer delight. This was perceived to have very positive effects to employee satisfaction and thus employee attitudes in terms of Figure 1.

"We do help some "begging" customers occasionally to get their stuff delivered next day, although they have actually ordered too late in the afternoon to get next day delivery. It felt quite nice that my boss [Transport Manager] actually said that the customer had thanked us because I was still able to get his goods into the truck. Never thanked me for that sort of thing before. It seemed to be important for the customer to get the ordered pipes next day, so I did a little extra work, because of the hassle the customer would have otherwise had at his construction site." (A transport co-ordinator, August 2002)

Strategic Level

At the strategic level, the logistics management became more interested in utilizing customer complaint information with the implementation of the construction and started to review and analyse complaint data regularly, on a monthly and annual basis.

> "Analysing the data seems to become more and more interesting as the size of the database grows. The database now has five months' data, and it's becoming quite interesting to play with the data in Access, and see whether anything new comes out. I can hardly wait until we can start looking at the data on an annual basis, when I expect we can better see the spread of different error types and can evaluate their cost effects and use that for improving our operations." (Logistics Manager, August 2002)

> "I have always believed that customer information is a key for achieving a new kind of, even strategic competitive advantage. However, I often wondered what is the relevant information we ought to get from the customer, and how we should go about getting that information. These customer satisfaction questionnaires we send out seem, from the perspective of improving logistics, rather useless. With regard to this new customer feedback construction, I was not convinced at the beginning that complaints information is a sensible source of customer information. I didn't believe that it generates enough data for a reliable and systematic analysis. However, as it seems now, our large volumes cause a large number of complaints, even though they are relatively few proportionally [given the total volume]. A careful handling of customer feedback creates a surprising amount of useful and interesting data." (Director of Logistics, November 2002)

By investigating and analysing the data in the long run, the top logistics management could identify trends and larger issues in the delivery process. This led to a further investigation concerning the potential for redesigning parts of the delivery process. For example, the logistics management initiated discussions with the transport partners aiming to reduce the number of errors occurring when goods are transported from the warehouse to the customer's premises.

> "I meet with all our transport partners once a month to evaluate and go through current issues and ponder about how to develop operations and cooperation. Before the latest meeting with one of the transporters, I filtered out from the system all the errors that, according to the system, they had caused and sent them the list a couple of days beforehand. It was quite a confusion and surprise for both of us, as we both claimed that we were innocent and that the other party was solely responsible for the mistakes. However, the uniform reporting of errors created an

intensive and productive dialogue. Already during the first meeting we found a problem spot in the delivery process, which we obviously decided to fix as quickly as possible. I am going to do the same thing with all of our transport partners."
(Transport Manager, October 2002)

Numerous small changes in the warehouse operations were implemented during the observed period. As the construction continues to be in use, more will be done every month. In principle these changes, if correctly implemented, should lead to better operations, improvement in quality and reduction in the number of complaints. While it was not within the scope of this study to measure the effects in absolute numbers, the management reported clear improvements on those occasions where changes were made based on the construction.

"Yes, the system is still actively used. It now forms an integrated part of our customer contact handling in the customer service centre." (Business Planning Manager, May 2006)

Conclusion

This paper addressed the issue of utilizing customer complaint information as a source for customer-focused process improvement, which was argued to direct process improvements to those activities that generate most value to the customer. A previous study by Johnston (2001) had shown that a well-handled customer complaint process positively correlates with process improvements (the Engineering route), customer satisfaction (the Customer Orientation route), employee attitudes (the Human Resource route) and, ultimately, company performance (Figure 1). However, the literature acknowledged that the problem of 'how' to achieve process improvements by utilizing customer complaints remained largely unsolved (Johnston and Mehra, 2002). The present paper set out to address this problem by creating a novel construction that—along the lines suggested by process-oriented management teachings (see e.g. Albright and Roth, 1994; Berry and Parasura man, 1997; Mizuno, 1992; Reichheld and Sasser, 1990)—recorded qualitative customer complaint information together with the complaint handler's interpretation, and allowed these to be processed systematically and linked to the rest of the company's information system. Hence, both the complaint itself and the complaint handler's tacit knowledge became usable as a managerial tool both at operational and strategic levels. The study thus demonstrated the usability of various process management teachings and their compatibility with customer-focused thinking when correctly employed.

The construction was implemented in a large Finnish technical wholesale enterprise and the implementation was studied as part of the constructive case study research method (e.g. Lukka, 2000, 2005). The results attained through the implementation clearly demonstrate that it is possible to achieve business process improvements by utilizing customer complaint data, thus supporting the argumentation in previous research which has raised the role of customer information in the improvement of business processes (e.g. Jones and Sasser, 1995, Kohli and Jaworski, 1990, Slater and Narver, 1994, 1996). Moreover, in terms of the Customer Orientation, Engineering and Human Resource routes depicted in Figure 1, the results of the case study showed that these are intertwined rather than independent of each other. The process improvements achieved became manifest and had managerial implications on three distinguishable levels of organization.

Firstly, by making the complaint-handling process more effective and staff-friendly, the construction showed a direct impact on the time and costs associated with handling the complaints. The complaint handlers also felt that their contribution to the company was finally recognized and their work became more valued by the management. Thus, in terms of the model in Figure 1, the Engineering route also impacted the Human Resource route via the process improvement having an effect on employee attitudes.

Secondly, enabling the operations management to track back each complaint to its source improved the management's ability to monitor performance and intervene to remove a problem if required. The customer complaint data was actively used to make minor adjustments and improvements within processes. These adjustments decreased the number of errors made and had a direct impact on costs, thus 'engineering' the processes and the Engineering route continuously to become more efficient.

Thirdly, the feedback data was also utilized in aid of strategic decision making regarding the long term development of warehouse operations and the network of transport partnerships. This had a direct impact on the Engineering route. Arguably the construction also improves customer satisfaction and thus the Customer Orientation route, although this was not explicitly demonstrated in our case. The developed construction did not really change the way the company deals with an individual customer recovery, with the exception of adding the possibility to give feedback via internet. However, as the customers complain and respective process improvements are made, customers do not experience similar errors in the future, which will reflect on their satisfaction in the longer term.

Based on the results of the case study, we propose extending the original model adopted from Johnston (2001) as illustrated in Figure 2. Thus, we propose that process improvements may also have indirect effects on company performance by positively impacting employee attitudes and customer satisfaction. The increased customer satisfaction and improved employee attitudes, in turn, are likely to have a positive impact on enhanced utilization of the customer complaint system and subsequent process improvement, thus creating a positive loop in the long run. The extended model could serve as a foundation for future research on this topic. For example, the constructs in the model could be operationalized and studied utilizing structural equation modelling in order to examine the imp act of the different routes to company performance. This is where the limits of the present study become apparent: we do not have any hard, quantitative data to measure the effectiveness or efficiencies gained via the construction numerically. The scope of the study was set to describe a construction, a means by which customer complaints can be linked to processes. The developed construction helps to identify the weak points or the points of failure within a company's existing processes and provides a tool for the management to identify and analyse these points. Measuring the actual effects of such process improvements remains a task for future research.

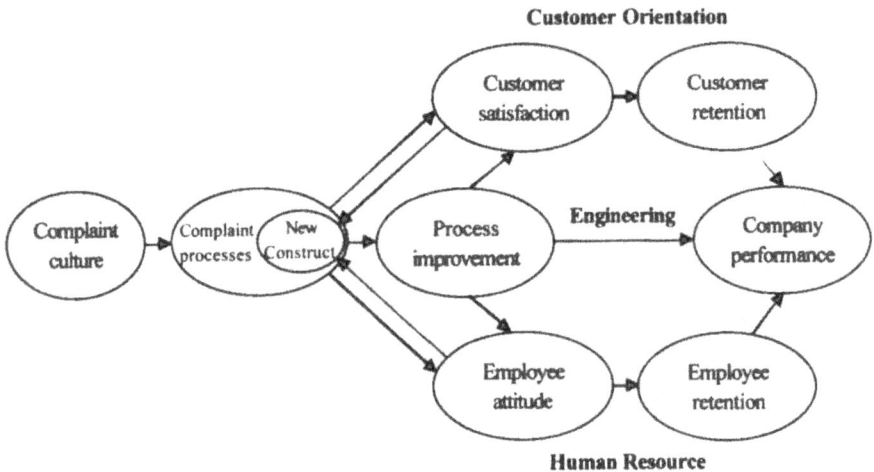

Figure 2: Process improvement as a central component of complaint management

A further limitation is imposed by the sectoral context of the study, which was set in the technical wholesale logistics environment and concentrated on the process of delivering the product from the shelves of the warehouse to the business customer. The complaints in this environment often relate to 'physical' faults, rather than the 'feeling' of service quality. The developed construction should be further developed to tackle 'softer' faults more common in true service industries. Further research on the applicability of this type of construction in other companies and other type of business environments would confirm its broader usability. However, in principle the constructive methodology does not require multiple cases due to its pragmatic nature. The fact that the construction was successfully implemented and still in use is relatively strong evidence that the construction works and is useful for managerial purposes. Finally, it is useful to point out that the construction is not 'automated' and that it cannot be taken out of its context, especially the company's organizational culture. The fact that the construction worked in HouseTech Corp was heavily dependent on the company already practicing process and quality management, and on the management of the company being committed to the project.

References

Albright, T.L., and Roth, H.P. (1994). Managing quality through the quality loss function. Journal of Cost Management, Winter, pp. 20–37.

Beer, S. (1985), Diagnosing the System for Organizations, Wiley, New York, NY.

Berry, L.L., and Parasuraman, A. (1997). Listening to the Customer—The Concept of a Service-Quality Information System. Sloan Management Review, Spring, pp. 65–76.

Bitner, M.J., Booms, B.H., and Tetreault, M.S. (1990). The service encounter: diagnosing favorable and unfavorable incidents. Journal of Marketing, 54, January, pp. 71–84.

Boshoff, C.R. (1997). An experimental study of service recovery options. International Journal of Service Industry Management, 8(2), pp. 110–130.

Boshoff, C.R. (1998). RECOVSAT: an instrument to measure satisfaction with transaction specific service recovery. Journal of Service Research, 1(3), pp. 236–49.

Brown, S.W., Cowles, D.L., and Tuten, T. (1996). Service recovery, its value and limitations as a retail strategy. International Journal of Service Industry Management, 7(5), pp. 32–46.

Cardy, R.L., and Dobbins, G.H. (1996). Human resource management in a total quality management environment: shifting from a traditional to a TQHRM approach. Journal of Quality Management, 1, pp. 5–20.

Checkland PB. (2000). Soft systems methodology: a thirty year retrospective. Systems Research and Behavioral Science 17: pp.11–58.

Clinton, B.D., and Hsu, K-C. (1997). JIT and the Balanced Scoreboard: Linking manufacturing control to management control. Management Accounting: Official Magazine of Institute of Management Accountants, Sep97, 79(3), p18.

Creech, B. (1994). The Five Pillars of TQM—How to Make Total Quality Management Work for You. Truman Talley Books/Plume, New York.

Davenport, T.H. (1993). Process Innovation: Reengineering Work Through Information Technology. Harvard Business School Press, Boston.

Deshpandé, R., Farley, J.U., and Webster, F.E.Jr. (1993). Corporate Culture, Customer Orientation, and Innovativeness in Japanese Firms: A Quadrad Analysis. Journal of Marketing, January, 57, pp. 23–27.

Earl, M., and Khan, B. (1994). How New is Business Process Redesign? European Management Journal, 12, No 1. pp. 20–30.

EFQM The European Foundation for Quality Management, 1997. What is Total Quality.

Feinberg, R.A., Widdows, R., Hirsch-Wyncott, M., and Trappey C. (1990), Myth and reality in customer service: good and bad service sometimes leads to repurchase, Journal of Consumer Satisfaction, Dissatisfaction and Complaining Behavior, 3, pp. 112–14.

Foster, G., Gupta, M., and Sjoblom, L. (1996). Customer profitability analysis: Challenges and new directions. Journal of Cost Management, Spring, pp. 5–17.

Hammer, M. (1990). Reengineering Work: Don't Automate, Obliterate. Harvard Business Review, July-August, pp. 104–112.

Hart, C.W.L., Heskett, J.L., and Sasser, W.E. (1990), The profitable art of service recovery, Harvard Business Review, July-August, pp. 148–56.

Järvinen, P., and Järvinen, A. (2004). Tutkimustyön metodeista. Opinpajan kirja, Tampere.

Jaworski, B., and Kohli, A. (1993). Market Orientation: Antecedents and Consequences. Journal of Marketing, July, pp. 53–70.

Johnston, R. (1995). Service failure and recovery: impact, attributes and process. Advances in Services Marketing and Management: Research and Practice, 4, pp. 211–28.

Johnston, R, (2001). Linking complaint management to profit. International Journal of Service Industry Management. 12(1), pp. 60–69.

Johnston, R., and Mehra, S. (2002). Best-practice complaint management. Academy of Management Executive, 16(4), pp. 145–154

Jones, T.O., and Sasser, W.E.jr. (1995). Why Satisfied Customers Defect? Harvard Business Review, November/December, pp. 88–89.

Kasanen, E., Lukka, K., and Siitonen, A. (1993). The Constructive Approach in Management Accounting Research, Journal of Management Accounting Research, Fall, 5, pp. 243–264.

Keating, P. (1995). A Framework for classifying and Evaluating the Theoretical Contributions of Case Research in Management Accounting. Journal of Management accounting research, Fall, pp. 67–

86. Knox, S. (1998). Loyalty-Based Segmentation and Customer Development Process. European Management Journal, 16(6), pp. 729–737.

Kohli, A.K., and Jaworski, B.J. (1990). Market Orientation: The Construct, Research Propositions, and Managerial Implications. Journal of Marketing, April, 54, pp. 1–18.

Lukka, K. (2000). The key issues of applying the constructive approach to field research, in Reponen, T. (ed.) Management Expertise for the New Millennium: In Commemoration of the 50th Anniversary of the Turku School of Economics and Business Administration. Publications of the Turku School of Economics and Business Administration, Series A-1:2000.

Lukka, K. (2003). Case study research in logistics. In Ojala, L. and Hilmola, O-P. (eds.) Publications of the Turku School of Economics and Business Administration, Series B 1: 2003, pp. 83–101.

Lukka, K. (2005). Approaches to Case Research in Management Accounting: The nature of empirical intervention and theory linkage. In Jönsson, S., and Mouritsen, J. (eds.), (2005). Accounting in Scandinavia—The Northern Lights. Liber and Copenhagen Business School Press, Kristianstad 2005.

Mizuno, S. (1992). Company-Wide Total Quality Control. Asian Productivity Organization, printing.

Moustakis, C. (1990). Heuristic Research. Design, Methodology and Applications. Newbury Park: Sage.

Oakland, J.S., and Oakland, S. (1998). The Links between people management, Customer satisfaction and business results. Total Quality Management, No 4&5, pp. 185–190.

Reichheld, F.F., and Sasser, W.E. jr. (1990). Zero Defections: Quality Comes to Services. Harvard Business Review, September/October, pp. 105–111.

Shank, J.K., and Govindarajan, V. (1994). Measuring the "cost of quality": A strategic cost management perspective. Journal of Cost Management, Summer, pp. 5–17.

Shapiro R., and Heskett J.L. (1985). Logistics Strategy, Cases and Concepts. West Publishing Company, USA.

Slater, S.F., and Narver, J.C. (1994). Does Competitive Environment Moderate the Market Orientatiton-Performance Relationship? Journal of Marketing, January, 58, pp. 46–55.

Slater, S.F., and Narver, J.C. (1996). Competitive Strategy in the Market-focused Business. Journal of Market Focused Management, 2, pp. 159–174.

Taylor, D. (1999). Supply-chain improvement: The lean approach. Logistics Focus, 7(1).

Turney, P.B.B. (1992). Activity-Based Management. Management Accounting— ABM puts ABC information to work. Management Accounting, January, 73(7), pp.20–25.

Wisner, A., and Kuorinka, I. (1988). Ergonomian menetelmäopin perusteita. Fysiologian ja psykologian käyttäminen työn todellisuuden tutkimisessa. In Scherrer, J.(ed.) Työn fysiologia. Porvoo: WSOY. pp. 591–610.

Vollmann, T., Berry, W., and Whybark, D. (1997). Manufacturing Planning and Control Systems. 4th edition. Irwin/McGraw-Hill, USA.

Wood, M. (1997). The notion of the customer in total quality management. Total Quality Management, 8(4), pp. 181–194.

Copyrights

Index

P

For Product Safety Concerns and Information please contact our EU
representative GPSR@taylorandfrancis.com
Taylor & Francis Verlag GmbH, Kaufingerstraße 24, 80331 München, Germany

www.ingramcontent.com/pod-product-compliance
Lightning Source LLC
Chambersburg PA
CBHW060815220326
41598CB00022B/2620

9 781774 632277